A Statistical
Manual
for Chemists

SECOND EDITION

A Statistical Manual for Chemists

Second Edition

Edward L. Bauer

Winthrop Laboratories
Rensselaer, New York

Academic Press **1971** New York and London

ACADEMIC PRESS, INC.
111 Fifth Avenue, New York, New York 10003

United Kingdom Edition published by
ACADEMIC PRESS, INC. (LONDON) LTD.
Berkeley Square House, London W1X 6BA

LIBRARY OF CONGRESS CATALOG CARD NUMBER: 73-154404

PRINTED IN THE UNITED STATES OF AMERICA

Contents

Preface to the Second Edition   ix

Preface to the First Edition   xi

List of Symbols   xiii

1   FUNDAMENTALS

1.1	Introduction	1
1.2	Experimental Error	2
1.3	The Average	2
1.4	The Normal Distribution	3
1.5	The t Distribution	5
1.6	Accuracy and Precision	6
1.7	The Average Deviation	7
1.8	The Variance and Standard Deviation	8
1.9	The Range	10
	References	12

2   THE AVERAGE

2.1	Replication	13
2.2	Confidence Limits	14
2.3	Degree of Confidence	15
2.4	Illustration of Confidence Limits	15
2.5	Calculation of Confidence Limits	16
2.6	Confidence Limits of Large Groups of Data by Range	19
2.7	Tolerance Limits	21
2.8	Invalid Measurements	22
2.9	Derivations and Proofs	23
	References	26

3   EXPERIMENTAL DESIGN AND THE ANALYSIS OF VARIANCE

3.1	Experimental Design	27
3.2	Nomenclature of Statistically Designed Experiments	29

3.3   Tests of Significance   30
3.4   The Analysis of Variance   31
3.5   Block Design: One-Way Classification   32
3.6   Block Design: Two-Way Classification   36
3.7   Models of ANOVA   42
3.8   Components of Variance   42
3.9   Expected Mean Square (EMS) Components   44
3.10 Latin Square   47
3.11 Factorial Experiments   49
3.12 Nested Factorial Experiment   55

4   THE COMPARISON OF TWO AVERAGES

4.1   The *t* Test   61
4.2   Uses of a *t* Test   62
4.3   Substitute *t* Tests   67
4.4   Uses of Substitute *t* Tests   68

5   ANALYSIS OF VARIANCE BY RANGE

5.1   Introduction   71
5.2   Block Design: One-Way Classification   72
5.3   Block Design: Two-Way Classification   75
5.4   Interaction   77
5.5   The Latin Square Design   80
5.6   Factorial Experiments   84
      References   93

6   CONTROL CHARTS

6.1   Introduction   95
6.2   Nomenclature   95
6.3   Theory of Control Charts   96
6.4   Control Limits   97
6.5   The Chart for Averages   98
6.6   The Chart for Ranges (or Standard Deviations)   98
6.7   Subgroups   98
6.8   Calculation of Control Limits   99
6.9   Significance of Control Limits   100
6.10 Runs   100
6.11 Making a Control Chart   101
6.12 Lack of Control   106
      References   108

7   CORRELATED VARIABLES

7.1	Linear Regression	109
7.2	A Laboratory Use of Regression	110
7.3	Shortcut Methods	117
7.4	Shortcut Method When X_n Does Not Equal nX_1	119
7.5	Colorimetric Analysis	120
7.6	Confidence Limits for X	126
7.7	Nonlinear Functions	127
	Reference	130

8   SAMPLING

8.1	The Sample and the Population	131
8.2	The Theory of Sampling	133
8.3	Sample Size	135
8.4	Attribute Sampling	136
8.5	Sampling by Variables	137
8.6	Use of Components of Variance	139
8.7	Variables Plan Based upon Normal Distribution	141
	References	144

9   CONTROL OF ROUTINE ANALYSIS

9.1	Problems of the Routine Analyst	145
9.2	Test for Outliers	146
9.3	Precision of the Analyses	147
9.4	Difference between Analysts	149
9.5	Accuracy	149
9.6	Precision of Optical Rotation Measurements	151
9.7	Precision of Colorimetric Analysis	152
9.8	Reduced Sample Size	153
9.9	Compliance with Specifications	156
9.10	Control Charts in the Analytical Laboratory	157
9.11	Interlaboratory Studies	158
	Reference	163

APPENDIX   165

Subject Index   191

Preface to the Second Edition

One of the cherished dreams of mankind is to be able to relive one's life, to have a second chance. Such a Faustian experience is granted to a few lucky souls, among whom are the authors of second editions. Given the chance to "do it over again," some bold persons would live an entirely different life, but most would probably try to keep the best of the old, risking the new and unknown only when the venture stood a good chance of producing an improvement.

This is what I have done. I still believe there is a need for a simple book on statistics for the working chemist, hence the purpose of the book (as set forth in the preface to the first edition) has remained the same.

In this edition, major changes have been made in every chapter except Chapter 1, and a chapter on control charts has been added.

Many of these changes were the result of letters received from readers of the first edition. I am very grateful to these people (most of them chemists) for their ideas and suggestions. I hope I have answered their questions and have in this way made this a more useful book.

Preface to the
First Edition

This book has been written for chemists who perform experiments, make measurements, and interpret data. Conversations with my colleagues have convinced me that too few chemists are taking advantage of the help statistical tools can give them: (1) maximum economy in experimentation, (2) maximum information from measurement data, (3) maximum accuracy and precision from test results. There seems to be a feeling that statistics are either too complicated to learn or too time consuming to use. My purpose has been to provide techniques which are both simple and fast and which will enable the chemist to analyze his own data.

The book is intentionally elementary in content and method. It is not meant to be a complete text on statistical techniques, but rather a manual for the working chemist. Throughout the book, use is made of rapid methods of calculation requiring only addition, subtraction, and the ability to use a slide rule. Discussions of statistical theory have been kept to a minimum because I believe many chemists are awed by a page of integrations, just as many mathematicians are abashed by a page of structural formulas.

In order to keep the manual simple and understandable to the neophyte, it was necessary to omit some very useful but sophisticated techniques. It is my hope that this manual will serve as a statistical primer; having mastered the fundamentals, the reader will be prepared to graduate to the more complex techniques as the need arises. At the end of each chapter the reader will find references, some of which are not cited in the text. These were included as a suggested step for further study.

I am grateful to Professor E. S. Pearson and *Biometrika* Trustees for permission to publish certain tables which appear in the Appendix.

I am also grateful to my secretary, Miss Pat Ibarreche, who so unflinchingly undertook the task of transcribing my original notes; to the Winthrop Laboratories' Librarian, Miss Ethel Center, for her tireless forages into the jungle of statistical literature for obscure references; and to Helen, my wife, for typing the manuscript, and for her criticism and coffee—both hot, strong, black, and without sugar.

List of Symbols

The reader often finds the terminology of a strange subject somewhat confusing. This is particularly true of statistical terminology where various authors use different symbols. The following glossary lists and defines the symbols used in this book, except in Chapter 6 which uses control chart symbols.

SYMBOL	DEFINITION
a	The intercept of a regression line
A	A factor that, when multiplied by the range, gives the confidence limits of the average
AD	Average deviation
α (alpha)	The risk of making a Type-I error
β (beta)	The risk of making a Type-II error
b	The slope of a regression line
c_1, c_2	Factors convert range to an unbiased estimate of the square root of the variance
CL	Confidence Limits of an average
d	A factor that converts average range to standard deviation
df	Degrees of freedom
f	Equivalent degrees of freedom
F	The critical value of the variance ratio test
I	A factor that when multiplied by the range gives the tolerance intervals of individual measurements
k	The number of groups in a series of observations

SYMBOL   **DEFINITION**

L	The critical value for the one-sample substitute t test based on range
M	The critical value for the two-sample substitute t test based on range
μ (mu)	The true mean of the population
n	Sample size or the number of observations in a group
N	kn—the total number of observations
p_1	The fraction defective of an acceptable lot
p_2	The fraction defective of an unacceptable lot
q	The critical value for the Studentized range
R	The range—the difference between the largest and the smallest of a group of observations
$\bar{R}$	The average range
s	The standard deviation of a sample
s_d	The standard deviation of a difference
$s_{\bar{x}}$	The standard deviation of an average $= s/\sqrt{n}$
σ (sigma)	The standard deviation of the population
σ^2	Variance of the population
Σ	Summation
t	The critical value of the t test
V	The variance of a sample
X	An observation, and the independent variable in a regression
$\bar{X}$	The average of a number of observations
$\bar{\bar{X}}$	The grand average
Y	The dependent variable in a regression

1

Fundamentals

1.1 INTRODUCTION

We use numbers in two ways—to enumerate objects and to designate the magnitude of measurements. If we were to count the number of words in the first sentence, we would find there are sixteen. No matter who counted them, when they were counted, or how they were counted, we would still get exactly sixteen words. This is an example of numbers used to enumerate objects. It is an absolute value—it does not change with time or method of measurement.

As a rule, chemists are not as interested in enumeration data as they are in measurement data. The information important to the chemist comes not from counting objects, but from weighing, reading burets, measuring volumes, and reading instruments. All of these operations involve measurements, and all measurements involve a region of uncertainty.

For example, consider the results obtained in reading the absorptivity of a spectrophotometric analysis. It is standard technique to make readings at about 0.43 absorbence to achieve minimum error. In this region, the absorbency scale of a well-known spectrophotometer is graduated so that there are 0.01 units between scale markings. The analyst must interpolate between 0.43 and 0.44, and the best he can hope to do is estimate one tenth of the least count of the instrument, or 0.001 units. This introduces a doubtful value into each reading. For instance, if the true absorbence of a solution is 0.435, and he reads the instrument scale as 0.436, he is making a relative error of 0.23 %; if he reads 0.433 or 0.437, the relative error is 0.46 %.

1.2 EXPERIMENTAL ERROR

Texts on analytic chemistry classify errors as determinate and indeterminate. Determinate errors are defined as those that can be avoided once they are recognized. This type of error is caused by such factors as:

(1) Improper calibration of glassware or instruments, or improper standardization of reagents.
(2) Personal errors, such as the tendency of an analyst to misjudge a color change.
(3) Prejudice.
(4) A constant error in method.

Determinate errors introduce a bias into the measurements. For example, if the analyst stands to one side of the hairline on the scale, his readings will all be high or low because of parallax.

Indeterminate errors cannot be eliminated. They exist by the very nature of measurement data. For example, the slight errors in interpolation are indeterminate. The analyst does not know their magnitude, or whether they are positive or negative. It is these indeterminate errors which we call "experimental error." They affect the precision of all chemical work, and we attempt to contain them in as narrow a zone as possible.

The most common way to minimize experimental error is to make a series of measurements on the same object and report the average.

1.3 THE AVERAGE

The average is the sum of the measurements divided by the number of measurements:

$$\bar{X} = (X_1 + X_2 + X_3 + \cdots + X_n)/n. \tag{1.1}$$

Two facts are evident from Eq. (1.1):

(1) The average is a measure of the central tendency; the sum of the deviations from it is zero.
(2) Since it comprises a number of different observations, it cannot be an absolute value.

The reliability of the average depends upon the range of values from which we obtain it: 16 is the average of 0 and 32; it is also the average of 15 and 17. As the average of 0 and 32 we could put little or no reliance upon its validity. As the average of 15 and 17, we can be reasonably certain that it is a good estimate of a true value. In neither case, can we feel as certain of the validity of the average as we can that there are sixteen words in the first sentence of this chapter.

When a chemist calculates an average, he is using statistics. He is intuitively making use of the laws of probability by taking advantage of the fact that he will make small errors more frequently than he will make large errors, and that in the long run, the plus-and-minus errors will cancel each other, leaving the average as a good estimate of the true value.

A rigorous definition of mathematical probability that would satisfy all statisticians would be too difficult and involved for this book. For our purpose, it is sufficient to describe, rather than define, it. Mathematical probability may be described as *expected frequency in the long run*.

To the statistician, the "long run" means a large number of data, variously distributed.

1.4 THE NORMAL DISTRIBUTION

Suppose the analyst makes a very large number of absorbence measurements (say 1000) on the same solution, and plots the magnitude of the measurement against the frequency of its occurrence. He would find the measurements distributed in a bell-shaped manner, with most of the measurements in the center and an equal number distributed with decreasing frequency on either side of the center. The distribution of such data can be described by a curve like Fig. 1.1, curve *a*. This is the Gaussian, or normal distribution curve. It is the theoretical distribution of the relative frequency of a large number of observations made on the same object. It is, therefore, a description of the expected frequency (or probability).

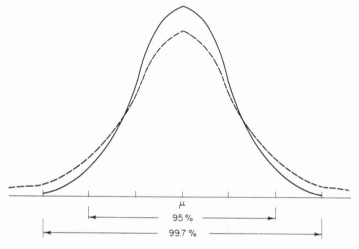

Figure 1.1. Curve a (——): Normal distribution curve. Curve b (– – –): Student distribution curve.

The curve has two properties that make it valuable to anyone using statistics:

(1)   It can be completely described by the average (μ), which fixes the location of the center of the curve with reference to the *x* axis, and by the standard deviation (σ), which describes the spread of the data along the *x* axis.

$$\sigma = [\Sigma(X - \mu)^2/n]^{1/2}. \tag{1.2}$$

(2)   The distribution of the frequency of the data has been thoroughly established. For example, 95 % of the individual measure-

ments will lie within $\mu - 1.96\sigma$ and $\mu + 1.96\sigma$, and a spread of $\mu \pm 3\sigma$ will include 99.7 % of the measurements.

This means that the analyst who is making 1000 absorbence readings would expect 50 readings to be outside $\mu \pm 1.96\sigma$ and only 3 readings to be outside the limits $\mu \pm 3\sigma$.

From a practical viewpoint, the converse is important. Suppose the analyst is making readings of a solution whose absorbence is 0.435, and he knows $\sigma = 0.005$. A reading outside of the range 0.425–0.445 would happen only 5 times in 100, and hence must be suspect. This concept is the basis of tests of significance.

1.5 THE *t* DISTRIBUTION

The theory of the normal distribution was developed from large amounts of data, and does not necessarily apply to small numbers of observations. In the laboratory, we cannot afford to make a very large number of observations; as a result, statistical tests based on the normal distribution could lead the laboratory worker to draw false conclusions. This fact was recognized by W. S. Gosset, an Irish chemist. In 1908, he published a paper under the pseudonym, "Student" entitled "The Probable Error of a Mean" (*1*). Partly by means of theoretical considerations, and partly by drawing small random samples, he derived the theoretical distribution of the average of small samples drawn from a normal distribution.

If we do not use large samples, we cannot know the true standard deviation σ or the true population mean μ. However, we can replace σ by the sample standard deviation (*s*). When we do this, we must use a new distribution, which is independent of σ. This is the concept introduced by Gosset that has become known as "Student's *t*"

$$t = (\bar{X} - \mu)/s_{\bar{X}}. \qquad (1.3)$$

Student demonstrated that the distribution of *t* is dependent only on the sample size (*n*). Figure 1.1, curve *b* (dashed line), shows the relationship of the *t* distribution to the normal distribution. The *t* curve is flatter than the normal curve, but approaches it as the sample size increases, becoming equal to the normal curve as n approaches infinity. For practical purposes, we usually use the normal distribution for sample sizes greater than 30.

It is necessary to understand the concept of the t distribution because it is the foundation upon which all tests of significance involving the comparison of two averages from small samples are based.

1.6 ACCURACY AND PRECISION

Accuracy may be defined as the correctness of a measurement. if

μ = the true value,

X = the value obtained experimentally,

E = the error,

then

$$\mu = X \pm E.$$

In chemical work, μ is often unknown, and therefore must be estimated from $X \pm E$. If E is zero, $\mu = X$, and the measurement is accurate.

Precision is a measure of the reproducibility of the measurements. The terms "accuracy" and "precision" are sometimes used interchangeably. They are not necessarily synonymous, as Fig. 1.2

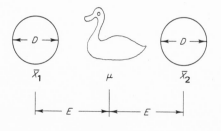

Figure 1.2

demonstrates. A hunter fired both barrels of a shotgun at a duck, with the results illustrated. Both barrels shot precisely, but the aim was not accurate—the duck flew away.

The true value is μ, the duck. The averages of the bursts with maximum distribution D are $\bar{X}_1$ and $\bar{X}_2$, a measure of the precision. The distance from $\bar{X}_1$ or $\bar{X}_2$ to μ, is E, a measure of the accuracy. It is only when E is small compared with D that accuracy and precision are the same.

If D were 10 ft, the hunter would miss the duck by $E - D/2 = 5$ ft. If, however, E were 5 ft, he would have a duck dinner.

Most statistical techniques measure precision rather than accuracy. However, statistical techniques are essential to the measurement of accuracy, because precision must be known before accuracy can be evaluated. A chemist cannot say a method is "accurate within the limits of experimental error" if he has no knowledge of the magnitude of the experimental error.

There are three common ways of evaluating the precision: (*i*) the average deviation, (*ii*) the variance, (*iii*) the range.

1.7 THE AVERAGE DEVIATION

If we sum all the X's in Fig. 1.3, regardless of whether they are

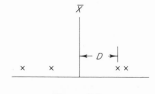

Figure 1.3

positive or negative, we will obtain the total deviation. Dividing this by the number of deviations will give the average deviation (AD).

$$AD = \Sigma(X - \bar{X})/n. \tag{1.4}$$

Example 1.1

Find the average deviation of the following moisture determinations: 0.48, 0.37, 0.47, 0.40, 0.43.

Step 1.   Find the average:

$$\bar{X} = \frac{0.48 + 0.37 + 0.47 + 0.40 + 0.43}{5} = 0.43.$$

Step 2.   Subtract $\bar{X}$ from each individual X and obtain the sum, disregarding the sign:

HOH (%)	$X - \bar{X}$
0.48	0.05
0.37	0.06
0.47	0.04
0.40	0.03
0.43	0.00
$\bar{X} = 0.43$	0.18

$$AD = 0.18/5 = 0.036$$

The average deviation is not an accurate measure of precision, because it gives a bias to the measurements, making them appear more precise than they really are (2).

1.8 THE VARIANCE AND STANDARD DEVIATION

Variance (V) is the sum of the squares of the deviations from the average divided by the degrees of freedom. The standard deviation is the square root of the variance.

$$V = \Sigma(X - \bar{X})^2/(n - 1); \qquad (1.5)$$

$$s = (V)^{1/2}$$

$$= [\Sigma(X - \bar{X})^2/(n - 1)]^{1/2}. \qquad (1.6)$$

We use $(n - 1)$ instead of n because we are using the sample average instead of the true mean. Statisticians have found that the sum of squares from the average of a sample is less than the sum of squares from the population average, and use of degrees of freedom (df) eliminates this bias. For example, if we have two observations, 2 and 4, from a population whose known average is 2, using Eq. (1.2),

$$\sigma^2 = [(2 - 2)^2 + (4 - 2)^2]/2 = 2.$$

If we use the sample average:

$$[2 + 4]/2 = 3,$$

and substitute in Eq. (1.2)

$$\sigma^2 = [(2 - 3)^2 + (4 - 3)^2]/2 = 1.$$

Using $df(2 - 1)$ as the divisor, as in Eq. (1.5),

$$V = [(2 - 3)^2 + (4 - 3)^2]/(2 - 1) = 2.$$

Standard deviation and variance are the most efficient measures of precision, and, as such, are the basis for all statistical tests.

Example 1.2

Calculate the variance and the standard deviation of the data in Example 1.1.

Step 1. Obtain the average of the moisture determinations: $\bar{X} = 0.43$.

Step 2. Subtract $\bar{X}$ from each individual X algebraically. The sum of the $(X - \bar{X})$ column should equal zero.

Step 3. Square each $(X - \bar{X})$ and obtain the sum. Steps 1, 2, and 3 are illustrated in the following tabulation:

HOH (%)	$(X - \bar{X})$	$(X - \bar{X})^2$
0.48	0.05	0.0025
0.37	−0.06	0.0036
0.47	0.04	0.0016
0.40	−0.03	0.0009
0.43	0.00	0.0000
$\bar{X} = 0.43$	$\Sigma = 0.00$	$\Sigma = 0.0086$

Step 4. Divide $\Sigma(X - \bar{X})^2$ by $n - 1$

$$V = \Sigma(X - \bar{X})^2/(n - 1)$$

$$= 0.0086/4$$

$$= 0.00215.$$

Step 5. Extract the square root of the variance. This is the standard deviation:

$$s = (0.00215)^{1/2}$$

$$= 0.046.$$

Recognizing the tedium of these calculations, statisticians have devised a substitute measure of precision—the range.

1.9 THE RANGE

The range is the difference between the highest and lowest value in a set of data. If we arrange the data in descending order

$$X_n \cdots X_{n-1} \cdots X_{n-2} \cdots X_1,$$

$$R = X_n - X_1. \tag{1.7}$$

Although it is not as efficient as the standard deviation, the ease of calculation has made it a popular technique for figuring precision. The relationship of the range to the standard deviation has been studied by Tippett (*3*), who has demonstrated that for small groups of measurements (the case usual in chemical laboratories) the range is accurate enough for practical purposes.

For this reason, range will be used in this book except where it is absolutely necessary to use variance, in which case, the variance will be calculated from the range. This will be done by using certain factors given in Tables V and VI of the Appendix.[1] The calculation of standard deviation from the range, and an introduction to the use of Table V, is illustrated by Example 1.3.

Example 1.3

Find the standard deviation from the range of the measurements in Example 1.1.

HOH (%)
0.48
0.37
0.47
0.40
0.43

Step 1. Substitute in Eq. (1.7):

$$R = 0.48 - 0.37 = 1.11.$$

Step 2. Substitute R as

$$s = R/c_1. \tag{1.8}$$

[1] Hereafter, when table numbers are given as roman numerals, it will be understood that we are referring to tables appearing in the Appendix, beginning on page 165.

In this example, we have one range of five numbers: $k = 1, n = 5$. We use Table V and look up the value for c_1 under the $k = 1$ column, $n = 5$ row. It is 2.48. This value is substituted in Eq. (1.8):

$$s = 0.11/2.48$$

$$= 0.044.$$

If we desire the variance:

$$V = s^2$$

$$= 0.001936.$$

REFERENCES

1. "Student" (1908). Collected Papers (2) (E. S. Pearson and John Wishart, eds.). *Biometrika* **6**, 1; Biometrika Office, University College, London, 1947.

2. Youden, W. J. (1950). Misuse of average deviation. *Nat. Bur. Standards (U.S.), Tech. News Bull.* **34**, 9.

3. Tippett, L. H. C. (1925). On the extreme individuals and range of samples taken from a normal population. *Biometrika* **17**, 364.

2

The Average

2.1 REPLICATION

The primary purpose of a chemical analysis is to estimate the true value μ of some property of material from a relatively small sample. Every analysis has an inherent variability arising from the usually small and unavoidable variations of manipulation, environment, and measurement. The precision of the analysis is limited by this variability and can be improved only by refining the method or by replication.

If the method has been refined to the best of the analyst's ability and equipment, the only way remaining to improve precision is by replication.

It is common practice for a chemist to run duplicate or triplicate analyses at the same time or on aliquots of the same solution. He then averages his results and judges the experimental error by the closeness of the two or three analyses. If the range of the results is small, he intuitively feels that he has an accurate and precise analysis. This is not necessarily true. For example, if the analyst were to weigh a sample, dissolve it and analyze aliquot portions of the solution, his results would probably check perfectly. However, this analysis would completely mask any small error in weighing, as well as any environmental error. If he were to make two weighings and simultaneously run through two analyses, he would have lessened his chance for masking weighing errors, but environmental and manipulation errors would be masked. Moran (*1*) demonstrated that duplicate analyses run at the same time were not truly independent: There is correlation between them. Powers (2) found that one chemist doing many analyses obtained better check results than did several chemists analyzing the same sample. Both these studies clearly demonstrate the importance of doing check analyses independently. In this manual, the term "replication" refers to a repetition of observations made independently.

If the analyst makes a second truly independent analysis, he will improve his knowledge in two ways:

(1)   The average of the two analyses will give a better estimate of the true value than will a single analysis.

(2)   The difference between the two measurements will give an estimate of the error; that is,

$$X_1 - X_2 = \mu + E_1 - (\mu + E_2)$$
$$= E_1 - E_2.$$

In effect, the analyst now has enough data for two estimates:

(1)   A "point estimate" $\bar{X}$ of the true value. This is the best unbiased estimate of μ available.

(2)   An "interval estimate," called a "confidence interval," within which he can assert that the true value of μ lies.

2.2 CONFIDENCE LIMITS

The confidence interval is calculated from the values of the sample analysis, (in this case, $X_1 - X_2$). It is an interval bounded by two values, called confidence limits, which are determined by (*i*) the sample size, (*ii*) the variability of the sample statistics, and (*iii*) the degree of confidence desired.

Let us examine the case of the analyst who has made duplicate analyses and has one result X_1 and a second result X_2, where $X_1 > X_2$. He can have 50% confidence that the true value lies between X_1 and X_2. Furthermore, he can have about 80% confidence that the true mean lies between $X_1 + R$ and $X_2 - R$, where $R = X_1 - X_2$, and about 95% confidence that μ lies between $X_1 + 6R$ and $X_2 - 6R$. This statement is proved in Section 2.9.3.

If the analyst makes a third independent analysis, he now has decreased his confidence interval considerably. As will be demonstrated at the end of the chapter, he has about 77% confidence that the range $X_1 - X_3$ ($X_1 > X_2 > X_3$) has a 77% chance of containing μ and 96% confidence that μ lies between $X_1 + R$ and $X_3 - R$ (Section 2.9.3).

2.3 DEGREE OF CONFIDENCE

Whenever we make a decision, we take a risk of making a wrong one. In statistical work, we can do this in two ways.

(1)   There is the risk, designated as α, of rejecting a good result.
(2)   There is the risk, designated as β, of accepting a bad result.

In making a statement about confidence limits we wish to avoid the α error a large percentage (say 90, 95, or 99%) of the time. Consequently, we make calculations that will keep α to say 0.1, 0.05, or 0.01.

2.4 ILLUSTRATION OF CONFIDENCE LIMITS

The following example is given to illustrate the calculation of confidence limits, and the effect of repeated replication:

Example 2.1

Suppose a chemist is analyzing a substance using a method of unknown variability. He obtains a result of 49.69. This is his estimate of the true value, but since he has no idea of the magnitude of experimental error, he has no idea of how much confidence he can put into this estimate.

He repeats the assay and obtains a result of 50.90. He now has another estimate of μ (50.30) and a range (50.90 − 49.69 = 1.21) from which to estimate the region within which the unknown true value lies. He can be 50% sure that the true value lies between 50.90 and 49.69 and about 80% confident that it lies between 52.11 and 48.48.

If he continues to repeat the analysis and obtains results of (3) 48.99, (4) 51.23, (5) 51.47, and (6) 48.80, he has improved his estimate of μ and the confidence he can place in this estimate. This can be seen from Table 2.1.

Table 2.1   The Effect of Replication

Analysis	X	$\bar{X}$	Range	95% Confidence limits
1	49.69	49.69	—	—
2	50.90	50.30	1.21	42.60–58.00
3	48.99	49.86	1.91	47.38–52.34
4	51.23	50.20	2.24	48.59–51.81
5	51.47	50.46	2.48	49.21–51.71
6	48.80	50.18	2.67	49.10–51.26

Note that although the range of these values increases, the confidence limits decrease with increasing sample size.

2.5 CALCULATION OF CONFIDENCE LIMITS

Confidence limits are expressed as the bounds of uncertainty about the average caused by the variability of the experiment.

Expressing the last sentence in the form of an equation, we have:

$$CL = \bar{X} \pm a \ \text{(variability)}, \tag{2.1}$$

where a is some constant, the value of which depends upon how the variability is expressed, the sample size, and the degree of confidence the experimenter desires.

If the variability is expressed as the sample standard deviation (s), Eq. (2.1) becomes

$$CL = \bar{X} \pm ts/\sqrt{n}, \tag{2.2}$$

where t is Student's t, values of which are given in Table XI.

If variability is expressed as range (R), Eq. (2.1) becomes

$$CL = \bar{X} \pm A\, \Sigma R, \tag{2.3}$$

where A is a factor derived from the relationship between the range and the standard deviation in a normal distribution, and from Student's t. The derivation of Eq. (2.3) is given in Section 2.9.

Example 2.2

Using the data in Example 2.1, we shall calculate the confidence limits by means of the standard deviation.

Step 1.   Calculate the average:

$\bar{X} = (49.69 + 50.90 + \cdots + 48.80)/6 = 50.18$.

Step 2.   Subtract $\bar{X}$ from each individual value of X; note that the sum of $(X - \bar{X}) = 0$.

Step 3.   Square each value of $(X - \bar{X})$.

Step 4.   Add $(X - \bar{X})^2$.
These steps give Table 2.2.

Step 5.   Calculate the variance,

$s^2 = \Sigma(X - \bar{X})^2/(n - 1) = 6.8456/5 = 1.3691$.

Step 6.   Calculate the standard deviation,

$s = (s^2)^{1/2} = (1.3691)^{1/2} = 1.17$.

For six observations ($n = 6$) we have 5 *df*. In Table XI, we find the value $t = 2.57$ for 5 *df*, for $\alpha = 0.05$.

Table 2.2

X	$(X - \bar{X})$	$(X - \bar{X})^2$
49.69	−0.49	0.2401
50.90	0.72	0.5184
48.99	−1.19	1.4161
51.23	1.05	1.1025
51.47	1.29	1.6641
48.80	−1.38	1.9044
$\bar{X} = 50.18$	$\Sigma = 0.00$	$\Sigma = 6.8456$

Step 7.   Calculate the confidence limits

$$CL = 50.18 \pm 2.57\,(1.17)/\sqrt{6}$$

$$= 48.95 - 51.41.$$

Example 2.3

Using the data in Example 2.1, we will calculate the 95 % confidence limits in Table 2.1 from the range.

(a)   For $n = 2$, $R = 1.21$. In Table I for $k = 1$, $n = 2$, and $\alpha = 0.05$, $A = 6.36$.

$CL = 50.30 \pm 6.36\,(1.21) = 42.60$–$58.00$.

(b)   For $n = 3$, $R = 1.91$. $A = 1.30$.

$CL = 49.86 \pm 1.30\,(1.91) = 47.38$–$52.34$.

(c)   For $n = 4$, $R = 2.24$. $A = 0.719$.

$CL = 50.20 \pm 0.719\,(2.24) = 48.59$–$51.81$.

(d)   For $n = 5$, $R = 2.48$. $A = 0.505$.

$CL = 50.46 \pm 0.505\,(2.48) = 49.21$–$51.71$.

(e) For $n = 6$, $R = 2.67$. $A = 0.402$.

$CL = 50.18 \pm 0.402 \ (2.67) = 49.10\text{--}51.26$.

These are the values given for the 95% confidence limits in Table 2.1. The reader will readily note that there is a difference between the confidence limits calculated by the two methods. The limits computed by using standard deviation are 0.30 units wider. These are the more accurate limits, since the standard deviation is the more precise way of calculating variation. However, the simplicity and ease of calculation of confidence limits from the range justifies its use.

2.6 CONFIDENCE LIMITS OF LARGE GROUPS OF DATA BY RANGE

In Section 1.9, we stated that, if the number of observations were small, range could be substituted for standard deviation without appreciable loss of accuracy. Range can still be used when the number of observations exceeds ten by breaking the data into k groups of n numbers, finding the total range of the k groups and using the A factor in Table I.

Example 2.4

Given the following observations on the same object, calculate the average and confidence limits.

33, 32, 30, 31, 22, 29, 32, 24, 34, 33, 33, 25, 34, 26, 29, 35, 33, 34.

Step 1. Here $n = 18$; therefore, divide the data into $k = 3$, $n = 6$, selecting the sequence at random. Such an arrangement might be:

32	33	29
30	31	32
24	33	34
25	34	26
34	22	29
33	33	35

$\bar{X} = 30.5$

Step 2.   Calculate the range of each subgroup:

R column $1 = 10$

R column $2 = 12$

R column $3 = 9$

$$R = 31$$

From Table I at $k = 3, n = 6$:

$A_{0.05} = 0.065,$

$A_{0.01} = 0.91,$

$95\% \, CL = 30.5 \pm 0.065 \, (31)$

$= 28.48$ to $32.52,$

$99\% \, CL = 30.5 \pm 0.091 \, (31)$

$= 27.68$ to $33.32.$

Example 2.4.1

If we desired to use the more precise standard deviation method, the calculations would be as given in the table on the opposite page.

$$s^2 = 256.50/(17) = 15.0882,$$

$$s = 3.90,$$

$$95\% \, CL = 30.5 \pm 2.11(3.90)/\sqrt{18}$$

$$= 28.56 \text{ to } 32.44,$$

$$99\% \, CL = 30.5 \pm 2.898(3.90)/\sqrt{18}$$

$$= 27.84 \text{ to } 33.16.$$

X	$X - \bar{X}$	$(X - \bar{X})^2$
32	1.5	2.25
30	−0.5	0.25
24	−6.5	42.25
25	−5.5	30.25
34	3.5	12.25
33	2.5	6.25
33	2.5	6.25
31	0.5	0.25
33	2.5	6.25
34	3.5	12.25
22	−8.5	72.25
33	2.5	6.25
29	−1.5	2.25
32	1.5	2.25
34	3.5	12.25
26	−4.5	20.25
29	−1.5	2.25
35	4.5	20.25
		256.50

2.7 TOLERANCE LIMITS

Confidence limits gave the experimenter an interval in which he could locate μ with some degree of confidence. Assuming a normal distribution, he can also estimate the interval and the limits within which the individuals of the population he is analyzing should fall. These are called tolerance limits and are calculated from Eq. (2.4) and (2.5).

$$TL = \bar{X} \pm ts, \qquad (2.4)$$

or

$$TL = \bar{X} \pm I\Sigma R, \qquad (2.5)$$

where t is Student's t, as given in Table I, and I is a factor derived from the relationship between the range and standard deviation and Student's t. Values for I are given in Table II.

Example 2.5

Estimate the tolerance interval for the population from which the data in Example 2.4 were drawn using (a) standard deviation and (b) range.

(a) $TI = \bar{X} \pm ts$:

$95\% \, TI = 30.5 \pm 2.11 \, (3.90)$

$\qquad = 22.27\text{--}38.73.$

$99\% \, TI = 30.5 \pm 2.898 \, (3.90)$

$\qquad = 19.20\text{--}41.80.$

(b) $TI = \bar{X} \pm I\Sigma R$:

$95\% \, TI = 30.5 \pm 0.276 \, (31)$

$\qquad = 21.94\text{--}39.06 \qquad (k = 3, \quad n = 6).$

$99\% \, TI = 30.5 \pm 0.386 \, (31)$

$\qquad = 18.53\text{--}42.47.$

2.8 INVALID MEASUREMENTS

It sometimes happens that in the course of taking a series of measurements, several observations are grouped close together while one is a maverick: It is seemingly out of line. The maverick observation may be truly a part of the parent distribution and hence valid, but on the other hand, it may have been caused by some unobserved source of error and should be discarded. There are several criteria for judging the validity of measurements. We illustrate one criterion in the following example.

Example 2.6

An analyst obtains the following results: (1) 93.3%, (2) 93.3%, (3) 93.4%, (4) 93.4%, (5) 93.3%, (6) 94.0%.

Is the sixth result (94.0%) valid?

Step 1. Calculate the average of the six results:

$\bar{X} = 93.45.$

Step 2. Calculate the range:

$$R = 94.0 - 93.3 = 0.7.$$

Step 3. Calculate the absolute difference of the suspected value from the average and divide by the range:

$$t_i = |X - \bar{X}|/R$$
$$= (94.0 - 93.45)/0.7$$
$$= 0.79.$$

Step 4. Compare with the critical values in Table X and discard if t_i is greater than the tabulated value.

Since 0.79 is greater than 0.76 (for $n = 6$), we can feel justified in judging 94 % to be a maverick and discarding it.

2.9 DERIVATIONS AND PROOFS

2.9.1 *Derivation of Factor A*

If we break the data into k subgroups of n numbers each, the total sample size is $N = kn$.

Confidence limits are given by

$$CL = \bar{X} \pm ts/\sqrt{n}. \qquad (2.6)$$

The standard deviation can be estimated from the range by the expression

$$s = \bar{R}/c_1, \qquad (2.7)$$

where c_1 is a constant that depends on the subgroup size $(n)_1$.

Substituting Eq. (2.7) in Eq. (2.6):

$$\frac{ts}{\sqrt{n}} = \frac{t\bar{R}/c_1}{\sqrt{N}}$$
$$= \frac{t\bar{R}}{c_1}\sqrt{N}$$
$$= \frac{t\bar{R}}{c_1}\sqrt{kn},$$

but $\bar{R} = R/k$, so

$$t\bar{R}/c_1 \sqrt{kn} = t/c_1 k \sqrt{kn} \, (\Sigma R)$$
$$= A\Sigma R,$$

where $t/c_1 k\sqrt{kn} = A$.

2.9.2 *Derivation of Factor I*

Tolerance limits are given by $\bar{\bar{X}} \pm ts$

$$s = \bar{R}/c_1,$$
$$TI = ts = t\bar{R}/c_1$$
$$= t(\Sigma R)/kc_1$$
$$= I(\Sigma R),$$

where $I = t/kc_1$

2.9.3 *Proof of Section 2.3*

Given two measurements, $X_1 > X_2$, the upper limit is $U = X_1 + aR$ and the lower limit is $L = X_2 - aR$, where a is a constant.

$$U - L = X_1 + aR - (X_2 - aR)$$
$$= X_1 - X_2 + 2aR$$
$$= R + 2aR$$
$$= (1 + 2a)R.$$

Also,

$$U - L = \bar{X} \pm AR$$
$$= \bar{X} + AR - (\bar{X} - AR),$$

but,

$$\bar{X} + AR - \bar{X} + AR = (1 + 2a)R,$$

hence,

$$2A = 1 + 2a,$$

$$A = t/kc_1\sqrt{kn}.$$

For two determinations $(n = 2)$ when $k = 1$

$$A = t/c_1\sqrt{n}$$

$$= t/1.414(1.414)$$

$$= t/2.$$

$$2A = 1 + 2a,$$

$$2t/2 = 1 + 2a,$$

$$t = 1 + 2a. \tag{2.8}$$

From Eq. (2.8) and the probability values of Student's t we can conclude the following:

When a equals	t equals	and the probability of being between $U - L$ is
0	1	0.5
1	3	0.795
2	5	0.874
3	7	0.910
4	9	0.930
5	11	0.942
6	13	0.951

For three determinations when $k = 1$

$$A = t/c_1 \sqrt{n}$$
$$= t/1.91(1.732)$$
$$= t/3.31.$$
$$2A = 1 + 2a,$$
$$2t/3.31 = 1 + 2a,$$
$$0.6046t = 1 + 2a,$$
$$t = (2a + 1)/0.6046. \qquad (2.9)$$

From Eq. (2.9) and the probability values of Student's t, we can conclude the following:

When a equals	t equals	and the probability of being between $U - L$ is
0	1.65	0.770
1	4.96	0.965

REFERENCES

1. Moran, R. F. (1943). Determination of the precision of analytical control methods. *Ind. Eng. Chem., Anal. Ed.* **15**, 361.
2. Powers, F. W. (1939). Accuracy and precision of micro-analytical determinations of carbon and hydrogen. Statistical study. *Ind. Eng. Chem., Anal. Ed.* **11**, 660.

3

Experimental Design and the Analysis of Variance

3.1 EXPERIMENTAL DESIGN

Man has been designing experiments ever since he began asking questions about the world around him; hence, the design of experiments is nothing new. He has developed a systematic plan for attacking problems, which has come to be known as the "scientific method." Essentially, this method consists of the following elements:

(1) A problem is stated and defined.
(2) A hypothesis is advanced to explain the problem.
(3) Data are collected.
(4) The hypothesis is tested against the data.
(5) The hypothesis is accepted or rejected on the basis of its agreement with the data.

Testing the hypothesis against the data and accepting or rejecting it are matters of judgment, and any way in which the judgment can be made objective, rather than subjective, will be of service.

A statistically designed experiment gives an estimate of error that can be used as a standard by which the results of the experiment can be measured. It is more efficient than the classic design in that it produces more usable information with less experimentation. An efficiently designed experiment should meet the following five criteria:

(1)   It should give an unbiased measure of the main effects.
(2)   It should provide an unbiased estimate of the variability of the main effects.
(3)   When necessary, the experiment should furnish information as to possible interaction between the main effects.
(4)   It should keep the unresolved experimental error as small as possible.
(5)   It should give an estimate of this error.

These criteria cannot be met by a haphazard collection of data; the experiment must be planned in advance to answer certain specific questions, some of which are

(1)   What is the purpose of the experiment?

The purpose should be clearly defined before the experiment is begun, and it should not be changed during the experimentation.

(2)   What time, equipment, and methods are available?

This question is mainly an economic one, and involves the price we must pay to obtain the data. It also considers the planning and scheduling of the experiment.

(3)   What are the main effects to be measured?

This question relates to the purpose of the experiment and, like it, should be clearly defined.

(4)   What prior information do we have about this or similar experiments?

Every experienced experimenter knows that a trip to the library can sometimes save many hours of work.

These are merely general questions in the broad framework of the scientific method. The remaining questions are those which make the experiment a statistically designed experiment:

(5)   Are the main effects likely to interact with each other, and if they do, is such interaction important?

Not all experiments will detect the existence of synergistic effects and therefore the answer to this question affects the design of the experiment.

(6)   What variability can we expect from the main effects and from the unresolved error?

Any prior knowledge as to expected variability is useful in setting up the levels at which the tests are to be run. For example, if it is known that the effect being measured may vary by $\pm 10\%$, setting the levels 5% apart will confuse any real effects with the experimental error.

(7)   What is the standard by which we intend to measure differences in effects?
(8)   At what level of significance are differences in effects to be considered meaningful?

Questions 7 and 8 are related, and their answers are important to the design. One feature of the statistically designed experiment is this: The experiment itself furnishes a standard by which the effects are to be measured. The answer to Question 8 depends upon the judgment of the experimenter. Both questions should be settled before the experiment is begun. This avoids wishful thinking from influencing the judgment of the experimenter.

3.2 NOMENCLATURE OF STATISTICALLY DESIGNED EXPERIMENTS

The statistically designed experiment was first used in agricultural experiments in which "blocks" or "plots" of ground were subjected to "treatments" to produce "yields" or "effects." Consequently, much of the nomenclature of experimental design may sound strange to a chemist. When we speak of "treatment" we mean some condition such as pressure or temperature has been imposed on the subject of the experiment.

3.3 TESTS OF SIGNIFICANCE

A feature of the statistically designed experiment, not inherent in the classic design, is the furnishing of statistical proof that a significant difference does or does not exist.

Statistical proof is not proof in the strict sense of the term, but is a very high probability that a given hypothesis is true or false. A statistical hypothesis is an assumption about the population being tested. The procedure for testing a hypothesis is:

(1)   State the hypothesis that a significant difference does not exist. Symbolically it is designated H_0.

(2)   State an alternate hypothesis (H_1).

(3)   Select the proper statistical test. This depends upon what the experimenter is comparing and what he wants to know.

(4)   Select the significance level of the test. The level is chosen arbitrarily by the experimenter and depends upon the risk he wishes to take of making a Type-1 (false rejection) error (symbol: α).

The significance level is the value of the test, where H_0 will be rejected $100\alpha\%$ of the time if H_0 is true. For example, at a value of $\alpha = 0.05$, a false rejection of the hypothesis would take place 5% of the time.

(5)   Run the test and collect the required data.

(6)   Apply the proper test of significance and accept or reject H_0.

Illustration 3.1

Suppose we were testing a new analytical method for nitrogen using a standard substance containing 10.0% nitrogen.

(1) H_0: $\mu = 10.0\%$. (If the method is correct, the average will equal 10% within the limits of experimental error.)

(2) H_1: $\mu < 10.0\%$. (If method is not correct average will be less than 10.)

(3)   Use the t test. $t = (\bar{X} - \mu)/(s/\sqrt{n})$

(4) $\alpha = 0.05$. (The analyst is willing to take a 5% risk of rejecting the method when it is correct.)

Critical value for the one-sided t test at $\alpha = 0.05$ is when $t = 2.35$ for 3 df.

(5) Make four determinations with the new method. Suppose the result obtained is 9.8% with a standard deviation of 0.2%.

(6) $t = (9.8 - 10.0)/(0.2/\sqrt{4}) = 2.0$. Since 2.0 is less than 2.35 we accept H_0.

Illustration 3.2

Suppose we wish to determine if two methods have the same precision. The analyst runs nine samples by each method.

(1) H_0: $\sigma_1^2 = \sigma_2^2$, i.e., the methods have equal variance.

(2) H_1: $\sigma_1^2 > \sigma_2^2$

(3) Use F ratio for test: $F = s_1^2/s_2^2$

(4) $\alpha = 0.05$

(5) Run the experiment and calculate s_1^2 and s_2^2.

Say $s_1^2 = 16$ $s_2^2 = 10$.

(6) $F = 16/10 = 1.6$. Since 1.6 is less than 3.44, accept H_0. (The critical value for $f_1 = 8, f_2 = 8$, in Table VIIIA, is 3.44.)

3.4 THE ANALYSIS OF VARIANCE

The information obtained from a statistically designed experiment can be analyzed by a method known as the "analysis of variance" or by its acronym ANOVA. It is a technique by which it is possible to isolate and estimate the separate variances contributing to the total variance of an experiment. It is then possible to test whether or not certain factors actually produce significantly different results in the variables tested.

In the next several sections of this chapter we will show how the analysis of variance is applied to several different types of experimental designs.

3.5 BLOCK DESIGN: ONE-WAY CLASSIFICATION

The simplest multiaverage design used to compare the averages of several replications is the block design. As mentioned before, it was originally used in agricultural experiments where blocks of land were subjected to different treatments. As used in the laboratory, the block becomes the replication of treatments. The difference between treatments is completely independent of any block differences.

We will use this simple design to derive the basic concepts of ANOVA.

Any measurement (which we shall call X_{ij}) obtained by the experimenter is made up of three components:

(1)   The true unknown value μ.
(2)   The effect of the treatment (T_i).
(3)   The experimental error (E). That is:

$$X_{ij} = \mu + T_i + E_{ij}. \tag{3.1}$$

The whole point of a designed experiment is to separate the effects of T and E, estimate their values and test their significance. By way of explanation, we shall synthesize some data and analyze it.

Example 3.1

Let us suppose the experiment is designed to determine the efficiency of four different methods of drying a substance. Each method is replicated five times. The data are given in Table 3.1 and were obtained by postulating that Treatment 1 removes all the water, Treatment 2 removes all but 0.1%, Treatment 3 all but 0.2%, and Treatment 4 all but 0.5%. The true water content is 1.8%. There is a random error added to the data.

This error has a variance (s^2) of 0.0305%, and a mean of zero.

The values in the table were built up from $X_{ij} = \mu + T + e$. For example, the result for Treatment 2, Analysis d was generated as follows:

$$X_{2d} = \mu + T_2 + e_{2d}$$
$$= 1.8 + (-0.1) + 0.3 = 2.0.$$

Table 3.1 Percentage of Water

Analysis	Treatment				
	1	2	3	4	
a	1.9	1.7	1.6	1.1	
b	1.7	1.6	1.7	1.3	
c	2.0	1.5	1.9	1.6	
d	1.6	2.0	1.6	1.1	
e	1.9	1.6	1.4	1.2	
Treatment total	9.1	8.4	8.2	6.3	32.0
Treatment average	1.82	1.68	1.64	1.26	1.60

The experimenter knows $X_{2d} = 2.0$ but, of course, does not know μ, T_2, or e_{2d}, which are parameters of the population under investigation. His statistics can be used to estimate the parameters. The grand average $\overline{\overline{X}} = 1.60$ is an estimate of μ. The difference between the treatment average and the grand average is an estimate of the treatment effect, and the difference between the observed value and the treatment average is an estimate of the error.

Equation (3.1) becomes

$$X_{ij} = \overline{\overline{X}} + (\overline{X}_i - \overline{\overline{X}}) + (X_{ij} - \overline{X}_i).$$

Transposing $\overline{\overline{X}}$

$$(X_{ij} - \overline{\overline{X}}) = (\overline{X}_i - \overline{\overline{X}}) + (X_{ij} - \overline{X}_i). \tag{3.2}$$

For X_{2d} as an example,

$$(2.00 - 1.60) = (1.68 - 1.60) + (2.0 - 1.68).$$

Equation (3.2) states that the deviation of any observation from the grand average is the sum of the treatment average from the grand average plus the deviation of the observation from the treatment average.

In order to analyze the whole experiment these deviations must be added. However, as in the case of the average deviation in Chapter 1, the sum of the deviations would be zero, and to avoid this they must be squared, which gives rise to the fundamental equation of the analysis of variance.

$$\text{Total sum } (X_{ij} - \bar{\bar{X}})^2 = \text{Total sum } (\bar{X}_{ij} - \bar{\bar{X}})^2$$
$$+ \text{Total sum } (X_{ij} - \bar{X}_j)^2. \quad (3.3)$$

This equation states that the total sum of squares of deviations from the grand average is equal to the sum of squares between treatment averages and the sum of squares of deviations within treatment averages. This basic principle holds for all types of ANOVA.

The null hypothesis of the experiment is that there is no difference between treatments.

The actual ANOVA of the data in Table 3.1 is carried out as follows:

(1)   Calculate the correction by squaring the grand total and dividing by the total number of analyses.

$$C = (32.0)^2/20 = 51.20$$

(2)   Square the individuals and subtract the correction. This is the total sum of squares.

$$\begin{aligned} SS_{tot} &= (1.9)^2 + (1.7)^2 + (2.0)^2 + \cdots + (1.1)^2 - 51.20 \\ &= 52.62 - 51.20 \\ &= 1.42. \end{aligned}$$

(3)   Square the sum of the columns (treatments) in Table 3.1, divide by the number of observations and subtract the correction. This is the sum of squares due to treatments.

$$\begin{aligned} SS_{treat} &= [(9.1)^2 + (8.4)^2 + (8.2)^2 + (6.3)^2]/5 - 51.20 \\ &= 52.06 - 51.20 \\ &= 0.86. \end{aligned}$$

(4)   Subtract sum of squares of treatments from the total sum of squares. This is the sum of squares due to analytical error—in this

case the variation between analyses in each treatment.

$$1.42 - 0.86 = 0.56.$$

(5)   Each of the sum of squares must be divided by the degrees of freedom associated with it. These are determined as shown below. If t stands for the number of treatments and r for the number of replications in each treatment

$$\text{Treatments:} \quad (t - 1) = 4 - 1 = 3;$$

$$\text{Error:} \quad t(r - 1) = 4(5 - 1) = 16;$$

$$\text{Total:} \quad [(tr) - 1] = [(4 \times 5) - 1] = 19.$$

(6)   Set up an ANOVA table

Source of variation	df	Sum squares	Mean square
Between treatments	3	0.86	0.2866
Within treatments (error)	16	0.56	0.035
Total	19	1.42	

The mean square is the estimated variance for each source of variation and is obtained by dividing sum of squares by the degrees of freedom.

(7)   The test of significance is equal to or greater than $F = F_{0.01}$ in Table VIII.

$$F = 0.2866/0.0350 = 8.19.$$

Table VIII gives tables of the F distribution, which is the distribution of the ratio of two variances. Table VIIIA gives critical values for $F_{0.05}$; these values would be exceeded by chance only five times in 100. Table VIIIB gives the $F_{0.01}$ critical values; values which would be exceeded only once in 100 times.

To find F in the tables, use the column for f_1 (degrees of freedom in the numerator) and the row for f_2 (degrees of freedom in the denominator).

From Table VIIIB for $f_1 = 3$, $f_2 = 16$, $F_{0.01} = 5.29$ which is smaller than the calculated $F = 8.19$. Hence, the null hypothesis is rejected and we can say that there is a high probability that the difference between treatments is significant.

3.6 BLOCK DESIGN: TWO-WAY CLASSIFICATION

To carry our illustration one step further let us synthesize a problem where we have three chemists each making a single analysis by three different methods of analysis.
The model is

$$X_{ij} = \mu + C_j + M_i + E_{ij},$$

and the hypotheses are

$$H_0: \quad C_1 = C_2 = C_3,$$

and

$$H_0: \quad M_1 = M_2 = M_3.$$

Example 3.2

Let us suppose the grand average is 10, and that Analyst A makes no error, Analyst B makes an error of -1, and Analyst C makes an error of $+1$. There is no error in Method 1, an error of -2 in Method 2, and an error of $+2$ in Method 3. In addition, there is the experimental error associated with every process. Suppose it is -0.2, -0.1, -0.1, -0.1, 0, $+0.1$, $+0.1$, $+0.1$, $+0.2$, distributed in a random fashion. The details of the synthesis of the data are shown in Table 3.2.

Table 3.3 shows the result of this synthesis. It is a table of the actual results obtained by the analysts.

The ANOVA proceeds as follows:

Table 3.2

Analyst	Method 1 $\bar{X}$	C	M	E	Method 2 $\bar{X}$	C	M	E	Method 3 $\bar{X}$	C	M	E
A	10	+0	+0	+0.1	10	+0	−2	−0.2	10	+0	+2	+0
B	10	−1	+0	−0.1	10	−1	−2	+0.1	10	−1	+2	−0.1
C	10	+1	+0	+0.1	10	+1	−2	+0.2	10	+1	+2	−0.1

Table 3.3

Analyst	Method 1	Method 2	Method 3	Analyst Sum	Analyst Average
A	10.1	7.8	12.0	29.9	9.97
B	8.9	7.1	10.9	26.9	8.97
C	11.1	9.2	12.9	33.2	11.07
Method sum	30.1	24.1	35.8	90.0	
Method average	10.03	8.03	11.93	10.0	

Step 1. Square the grand sum and divide by the number of observations making up this sum. This is the correction factor.

$$C = (90)^2/9 = 900$$

Step 2. Square the individuals and subtract the correction. This is the total sum of squares.

$$SS_{tot} = (10.1)^2 + (8.9)^2 + \cdots + (12.9)^2 - 900 = 29.54$$

Step 3.   Square the sum of the columns, divide by the number of observations, and subtract the correction. This is the sum of squares due to methods.

$$SS_{meth} = [(30.1)^2 + (24.1)^2 + (35.8)^2]/3 - 900 = 22.82$$

Step 4.   Square the sum of the rows, divide by three, and subtract the correction. This gives the sum of squares due to analysts.

$$SS_{anal} = [(29.9)^2 + (26.9)^2 + (33.2)^2]/3 - 900 = 6.62$$

These values are placed in the ANOVA table (see Table 3.4).

Step 5.   The degrees of freedom for each variance are figured as follows:

$$\text{Methods} = (m - 1) = 3 - 1 = 2,$$
$$\text{Analysts} = (a - 1) = 3 - 1 = 2,$$
$$\text{Error} = (m - 1)(a - 1) = 2 \times 2 = 4,$$
$$\text{Total} = (ma - 1) = (3 \times 3) - 1 = 8.$$

Step 6.   Set up ANOVA Table 3.4.

Table 3.4   ANOVA Table

Source of variance	df	Sum squares	Mean square
Between methods	2	22.82	11.41
Between analysts	2	6.62	3.31
Error	4	0.10	0.025
	8	29.54	

The tests of significance are

(1)   Between analysts: $F = 3.31/0.025 = 132.40$
(2)   Between methods: $F = 11.41/0.025 = 456.40$

On consulting tables of the F distribution (Table VIII), we find that both are highly significant. The "between analysts" mean square has 2 df, and the error mean square has 4 df. Consulting Table VIIIB in the column $f_1 = 2$ and the row $f_2 = 4$, we find the value for $F_{0.01} = 18.0$. The calculated value for F (132.4) exceeds this critical value, so we can say the difference is significant with a probability of being wrong only once in 100 times. The same reasoning applies to the calculated value for F for "between methods," which is 456.4.

Example 3.3. *Interaction*

If Analyst A commits a systematic error of $+0.5$ with Method 3, and Analyst B commits a systematic error of -0.5 with Method 3, the linearity of the model would be destroyed, and we would be unable to uncover this interaction. Instead, we would have to design the experiment to follow the model

$$X_{ijk} = \mu + C_i + M_j + I_{ij} + E_{k(ij)}.$$

Assuming the interactions and the same distribution of experimental error as Example 3.2, and again assigning this distribution at random, we obtain Table 3.5.

Table 3.5

Replicates	Method 1 Analyst			Method 2 Analyst			Method 3 Analyst		
	A	B	C	A	B	C	A	B	C
1	10.1	8.9	11.1	7.8	7.1	9.2	12.5	10.9	12.4
2	9.8	9.0	10.9	7.9	7.2	9.1	12.4	11.1	12.6
Sum analyst	19.9	17.9	22.0	15.7	14.3	18.3	24.9	22.0	25.0
Sum (method)	59.8			48.3			71.9		
Grand sum	180.0								

This design is analyzed by setting up Table 3.6, which is a sum of methods and analysts table. For example, the number 19.9 under Method 1, Analyst A, is the sum of the two analyses performed by Analyst A by Method 1 (10.1 + 9.8).

The variances of Table 3.6 are analyzed by the ANOVA following the method outlined in Example 3.2; but because each figure is the sum of two, we divide by twice the regular divisor.

Step 1. Correction factor
$$(180)^2/(9 \times 2) = 1800.$$

Step 2. Analysts mean square
$$SS_{anal} = [(60.5)^2 + (54.2)^2 + (65.3)^2]/(3 \times 2) - 1800$$
$$= 10.33.$$

Step 3. Methods mean square
$$SS_{meth} = [(59.8)^2 + (48.3)^2 + (71.9)^2]/(3 \times 2) - 1800$$
$$= 46.42.$$

Step 4. Total variance of interaction: Sum the squares of the individuals in Table 3.6 and divide by 2
$$[(19.9)^2 + (15.7)^2 + \cdots + (25.0)^2]/2 - 1800 = 57.56.$$

Step 5. Methods–analysts interaction. This is the difference between the total variance of the interaction and the variances due to methods and analysts
$$57.56 - 10.33 - 46.42 = 0.90.$$

Step 6. Total variance. Sum the squares of the numbers in Table 3.5
$$(10.1)^2 + (9.8)^2 + \cdots + (12.6)^2 - 1800 = 57.69.$$

Step 7. Error variance
$$57.69 - 57.56 = 0.13.$$

Step 8. Degrees of freedom. The degrees of freedom for the main effects and the interaction are figured as they were in Example 3.2. The degrees of freedom for the error variance are $ma(r - 1)$, and the total degrees of freedom are $mra - 1$.

Table 3.6   Methods–Analysts Interaction

Analyst	Methods			Total
	1	2	3	
A	19.9	15.7	24.9	60.5
B	17.9	14.3	22.0	54.2
C	22.0	18.3	25.0	65.3
	59.8	48.3	71.9	180.0

Methods: $(m - 1) = 3 - 1 = 2$;

Analysts: $(a - 1) = 3 - 1 = 2$;

M × A: $(m - 1)(a - 1) = (3 - 1)(3 - 1) = 4$;

Error: $ma(r - 1) = 3 \times 3(2 - 1) = 9$;

Total: $(3 \times 3 \times 2) - 1 = 17$.

Step 9.   These values are substituted in Table 3.7, the ANOVA table.

Table 3.7

Source	df	SS^a	MS^a	EMS^b
Between methods	2	46.42	23.21	$V_e + 6V_m$
Between analysts	2	10.33	5.16	$V_e + 6V_a$
Interaction (M × A)	4	0.90	0.225	$V_e + 2V_{ma}$
Error	9	0.13	0.0144	V_e

[a] SS: Sum squares; MS: Mean square.

[b] Expected Mean Square.

Step 10.   Tests of significance

    1.   Interaction: $F = (V_e + 2V_{ma})/V_e = 15.6.$

    2.   Analysts: $F = (V_e + 6V_a)/V_e = 358.$

    3.   Methods: $F = (V_e + 6V_m)/V_e = 1612.$

    All are highly significant.

3.7 MODELS OF ANOVA

There are three kinds of ANOVA models—fixed, random, and mixed—depending upon the source of the data and the use to which the models are to be put.

1.   A fixed model is one in which all possible effects are investigated. The models discussed in Examples 3.5 and 3.6 were fixed models, because we were investigating the effects of the only three chemists available and the only three methods of analysis.

2.   A random model is one in which a sample of possible effects is analyzed. If in Example 3.5 we selected any three chemists as a sample of chemists in general, this would be a random model.

3.   A mixed model is one in which some of the effects are random and some are fixed. If in Example 3.6 we selected three chemists from several, the chemists effect is random and the method effect is fixed and we therefore have a mixed model.

Table 3.7 would now read as Table 3.8 (opposite).

In general, when the ANOVA is used to analyze existing data, it is a fixed model. When it is used as a method of statistical inference, it is either a random or mixed model.

3.8 COMPONENTS OF VARIANCE

The mean square column in the ANOVA table gives only an estimation of the population variance. Tables 3.5 and 3.6 have an extra column labeled EMS (expected mean square). This column gives the components of variance of the expected mean square.

Table 3.8

Source	df	SS	MS	EMS
Between methods	2	46.42	23.21	$V_e + 2V_{ma} + 6V_m$
Between analysts	2	10.33	5.16	$V_e + 6V_a$
Interaction (M × A)	4	0.90	0.225	$V_e + 2V_{ma}$
Error	9	0.13	0.0144	V_e

The EMS column serves two uses:

(1) It determines what mean square ratio to use for the F test. For example, in Table 3.7:

$$F \text{(analysts)} = (V_e + 6V_a)/V_e$$
$$= 5.16/0.0144$$
$$= 358$$

but

$$F \text{(methods)} = (V_e + 2V_{ma} + 6V_m)/(V_e + 2V_{ma})$$
$$= 23.21/0.225$$
$$= 103.2,$$

whereas, in Table 3.6:

$$F \text{(methods)} = (V_e + 6V_m)/V_e$$
$$= 23.21/0.0144$$
$$= 1612.$$

(2) It is used to estimate the magnitude of the variances. In Table 3.7:

$$0.2250 = V_e + 2V_{ma};$$
$$2V_{ma} = 0.2250 - 0.0144;$$
$$V_{ma} = 0.1053.$$

This is an estimate of the variance of the methods–analysts interaction. Similarly, the variance between the three analysts can be estimated:

$$6V_a = 5.1600 - 0.0144,$$

$$V_a = 0.8576.$$

The variance between methods is:

$$6V_m = 23.2100 - 0.225,$$

$$V_m = 3.8308.$$

3.9 EXPECTED MEAN SQUARE (EMS) COMPONENTS

Step 1.   Designate the number of levels in each main effect by some convenient letter. For instance, in Example 3.2 if we designate methods by m, analysts by a, and replications by r, numerically: $m = 3$, $a = 3$, and $r = 2$.

Step 2.   Designate the variances by V_m for variance between methods, V_a for variance between analysts, V_{ma} for the interaction variance, V_e for the error variance.

Step 3.   Set up the following table:

Mean square	EMS
Methods	$V_e + rV_{ma} + arV_m$
Analysts	$V_e + rV_{ma} + mrV_a$
M × A	$V_e + rV_{ma}$
Error	V_e

This is a table of all possible components.

Step 4.   If the model is a random model, all interaction components are used.

For a fixed model, cancel all interaction components in the main effects, but not in the interactions.

For a mixed model, observe the following rules:

(1)   When the one factor is fixed, the EMS of that factor contains all possible interactions.

(2)   The EMS of the random factors contains only the interactions of the random factors.

(3)   When two factors are fixed, cancel the terms of the two fixed factors interaction, and leave the fixed-random factor interaction terms.

(4)   The random factors contain no interaction terms.

To illustrate the above rules, suppose we have a three-level analysis with main effects G, T, S. We set up our table of possible EMS:

Mean square	Possible EMS
G	$V_e + sV_{gt} + tV_{gs} + tsV_g$
T	$V_e + sV_{gt} + gV_{ts} + sgV_t$
S	$V_e + tV_{gs} + gV_{ts} + gtV_s$
GT	$V_e + sV_{gt}$
GS	$V_e + tV_{gs}$
ST	$V_e + gV_{st}$

This is the proper table if all the components are random.

If all the components are fixed, we cancel the interaction terms and the table becomes:

Mean square	EMS
G	$V_e + ts V_g$
T	$V_e + sg V_t$
S	$V_e + gt V_s$
GT	$V_e + s V_{gt}$
GS	$V_e + t V_{gs}$
ST	$V_e + g V_{st}$
Error	V_e

If factor G is fixed, and factors T and S are random, we follow Rules 1 and 2 of Step 4;

Mean square	EMS	Rule
G	$V_e + s V_{gt} + t V_{gs} + ts V_g$	1
T	$V_e + g V_{ts} + gs V_t$	2
S	$V_e + g V_{ts} + gt V_s$	2

If factors G and T are fixed, and S is random, we carry out Rules 3 and 4 of Step 4:

Mean square	EMS	Rule
G	$V_e + t V_{gs} + ts V_g$	3
T	$V_e + g V_{st} + gs V_t$	3
S	$V_e + gt V_s$	4

In all cases, the interaction EMS terms remain the same.

3.10 LATIN SQUARE

The Latin square is a design in which at least three factors can be varied during the experiment. As the name implies, it is a square with n rows and n columns. Each of the treatments to be studied is replicated n times. These are distributed at random over the square so that each treatment occurs only once in each row and each column.

If we wish to investigate the effect of pressure, temperature, and catalyst on a reaction we could carry out nine experiments designated by one of several possible Latin squares. Such a square might be:

Table 3.9

	T_1	T_2	T_3
P_1	C_1	C_2	C_3
P_2	C_2	C_3	C_1
P_3	C_3	C_1	C_2

The symbols P_1, C_1, T_1 mean that all three factors are at the lowest level; P_2, C_2, T_2 mean the factors are at the second levels, and P_3, C_3, T_3 mean the factors are at the highest levels.

It is evident from Table 3.9 that each level of the catalyst appears only once in each row and each column.

The model is:

$$X_{ijk} = \mu + P_i + T_j + C_k + E_{ijk}.$$

The ANOVA will separate the sums of squares and hence the variance due to pressure, temperature, catalyst, and error.

Example 3.4

Substituting fictitious values in Table 3.9 we obtain Table 3.10. We will judge the differences to be significant at the 0.05 level.

The ANOVA is carried out as follows:

Step 1.　$SS_{\text{cor}} = (19.8)^2/9 = 43.56.$

Table 3.10

	T_1	T_2	T_3	Sum T
P_1	1.1	1.7	2.4	5.2
P_2	1.9	2.4	2.5	6.8
P_3	2.2	2.6	3.0	7.8
Sum P	5.2	6.7	7.9	
Sum C	$C_1 = 6.2$	$C_2 = 6.6$	$C_3 = 7.0$	
Grand sum	19.8			

Step 2. $SS_{ind} = (1.1)^2 + (1.9)^2 + \cdots + (3.0)^2 - 43.56 = 2.52.$

Step 3. $SS_{pres} = \{\frac{1}{3}[(5.2)^2 + (6.8)^2 + (7.8)^2]\} - 43.56 = 1.1467.$

Step 4. $SS_{time} = \{\frac{1}{3}[(5.2)^2 + (6.7)^2 + (7.9)^2]\} - 43.56 = 1.22.$

Step 5. $SS_{cat} = \{\frac{1}{3}[(6.2)^2 + (6.6)^2 + (7.0)^2]\} - 43.56 = 0.1067.$

Step 6. $SS_{err} = 2.52 - (1.1467 + 1.2200 + 0.1067) = 0.0466.$

Step 7. The degrees of freedom are: for pressure $(n - 1) = 2$; for temperature $(n - 1) = 2$; for catalyst $(n - 1) = 2$; for error $(n - 1)(n - 2) = 2$.

Step 8. Set up ANOVA table:

Source of variance	df	Sum squares	Mean square
Pressure	2	1.1467	0.5735
Temperature	2	1.2200	0.6100
Catalyst	2	0.1067	0.0534
Error	2	0.0466	0.0233

Step 9. Tests of significance:

catalyst $\qquad$ $0.0534/0.0233 = 2.29$;

temperature $\qquad$ $0.6100/0.0233 = 26.2$;

pressure $\qquad$ $0.5735/0.0233 = 24.6$;

$F_{0.05}$ $\qquad$ for $(f_1 = 2, \; f_2 = 2) = 19.0$,

$F_{0.01}$ $\qquad$ for $(f_1 = 2, \; f_2 = 2) = 99.0$.

Step 10. Conclusion: Since the $\alpha = 0.05$ level of significance was chosen, this experiment indicates that both temperature and pressure have a significant influence on the reaction. Note, however, that if the $\alpha = 0.01$ level had been chosen, we would not have found any significant difference. As it is, the experimenter is taking five chances out of 100 that the results of this experiment are giving him false information.

3.11 FACTORIAL EXPERIMENTS

An experiment designed so that several factors can be investigated at several levels is known as a factorial experiment. For instance, we may wish to study the effects of temperature, pressure, and catalyst at two levels each. We have to run $2^3 = 8$ experiments. From these eight experiments we can obtain information about

(1) The variation between temperature levels,
(2) The variation between pressure levels,
(3) The variation between catalyst levels,
(4) Interactions between temperature and pressure, temperature and catalyst, and catalyst and pressure,
(5) The temperature–catalyst–pressure interaction which can be used as an estimate of the experimental error.

Factorial experiments are comprehensive and efficient, but the number of experiments increases rapidly as the number of factors and levels increases. Four factors at two levels require $2^4 = 16$ experiments, and four factors at three levels require $3^4 = 81$ experiments.

Example 3.5

Suppose we have three lots of material on which we wish to try three methods of analysis. We also wish to determine if there is a difference between two analysts. Each analyst performs an analysis on each lot, using each method. We have $3 \times 3 \times 2 = 18$ experiments, and the following possible sources of variation:

(1)   Main effects between lots of material (L)
(2)   Main effects between methods of analysis (M)
(3)   Main effects between analysts (A)
(4)   Possible interactions between lots and methods (L × M)
(5)   Possible interactions between lots and analysts (L × A)
(6)   Possible interactions between methods and analysts (M × A)
(7)   Possible interactions between lots, methods, and analysts (L × M × A)

We shall use the L × M × A interaction for an estimate of error. The results of such a study are given in Table 3.11

Table 3.11

Analyst	Lot A Method			Lot B Method			Lot C Method		
	1	2	3	1	2	3	1	2	3
a	1.9	1.9	2.0	1.8	1.1	1.7	1.4	1.3	1.1
b	1.8	1.7	1.4	1.6	1.4	1.0	1.4	1.1	0.8
Sum columns	3.7	3.6	3.4	3.4	2.5	2.7	2.8	2.4	1.9
Sum lots	10.7			8.6			7.1		
Sum methods	(1) 9.9			(2) 8.5			(3) 8.0		
Sum analysts	(a) 14.2			(b) 12.2					
Grand sum	26.4								

The ANOVA is carried out as follows:

Step 1.   Calculate the correction

$$C = (26.4)^2/18 = 38.72.$$

Step 2.   Calculate total variance

$$SS_{ind} = (1.9)^2 + (1.8)^2 + (1.9)^2 + \cdots + (0.8)^2 - 38.72$$
$$= 2.12.$$

Step 3.   Set up interaction tables and analyze them.

Lots × Methods: Sum the method response for each Lot and enter in Table 3.12. For example, Lot A, Method 1 = 1.9 + 1.8 = 3.7; Lot A, Method 2 = 1.9 + 1.7 = 3.6, etc.

Table 3.12   Lots × Methods

Method	A	B	C	Sum methods
		Lot		
1	3.7	3.4	2.8	9.9
2	3.6	2.5	2.4	8.5
3	3.4	2.7	1.9	8.0
Sum lots	10.7	8.6	7.1	26.4

The ANOVA of the data in Table 3.12 proceeds as usual except the sums of squares of the numbers is divided by the number of individuals making up the number. That is to say, the sum square of individuals is:

$$SS_{tot} = \{\tfrac{1}{2}[(3.7)^2 + (3.6)^2 + \cdots + (1.9)^2]\} - 38.72$$
$$= 1.5400.$$
$$SS_{lots} = \{\tfrac{1}{6}[(10.7)^2 + (8.6)^2 + (7.1)^2]\} - 38.72 = 1.0900.$$
$$SS_{meth} = \{\tfrac{1}{6}[(9.9)^2 + (8.5)^2 + (8.0)^2]\} - 38.72 = 0.3273.$$

The L × M ANOVA table is Table 3.13.

Table 3.13 L × M ANOVA

Source	df	SS	MS
Between lots	2	1.0900	0.545
Between methods	2	0.3233	0.1617
L × M Interaction	4	0.1267	0.0317
	8	1.5400	

Lots × Analysts: Sum the lots response for each analyst as was done to make Table 3.12. This gives Table 3.14.

Table 3.14 Lots × Analysts

Analyst	Lot			Sum analysts
	A	B	C	
a	5.8	4.6	3.8	14.2
b	4.9	4.0	3.3	12.2
	10.7	8.6	7.1	26.4

Step 4. Tests of significance.

$M \times A$ interaction: $F = 0.1106/0.0305 = 3.63$, $\quad F_{0.05} = 6.94$;

$L \times A$ interaction obviously not significant;

$L \times M$ interaction obviously not significant.

Table 3.15   Methods × Analysts

Analyst	Methods			Sum analysts
	1	2	3	
a	5.1	4.3	4.8	14.2
b	4.8	4.2	3.2	12.2
	9.9	8.5	8.0	26.4

Table 3.16   L × A ANOVA

Source	df	SS	MS
Between lots	2	1.0900	0.545
Between analysts	1	0.2222	0.2222
L × A Interaction	2	0.0145	0.0072
	5	1.3267	

Table 3.17   M × A ANOVA

Source	df	SS	MS
Between methods	2	0.3233	0.1617
Between analysts	1	0.2222	0.2222
M × A Interaction	2	0.2212	0.1106
	5	0.7667	

Table 3.18

Source	df	SS	MS	EMS (fixed model)
Between lots	2	1.0900	0.5450	$V_e + 6V_l$
Between methods	2	0.3233	0.1617	$V_e + 6V_m$
Between analysts	1	0.2222	0.2222	$V_e + 9V_a$
L × M Interaction	4	0.1267	0.0317	$V_e + 2V_{lm}$
L × A Interaction	2	0.0145	0.0072	$V_e + 3V_{la}$
M × A Interaction	2	0.2212	0.1106	$V_e + 3V_{ma}$
L × M × A Interaction	4	0.1221	0.0305	V_e
	17	2.1200		

Since they are not significant, the interactions can be combined to form a new error term:

$(0.1267 + 0.0145 + 0.2212 + 0.1221)/(4 + 2 + 2 + 4)$
    $= V_e'.$

$V_e' = 0.0404$ with   12 *df.*

Between analysts,

$F = 0.2222/0.0404 = 5.50,$ $F_{0.05} = 4.75.$

Between methods,

$F = 0.1617/0.0404 = 4.00,$ $F_{0.05} = 3.89.$

Between lots,

$F = 0.5450/0.0404 = 13.5,$ $F_{0.05} = 3.89.$

Conclusions: The differences between lots, methods, and analysts are significant.

3.12 NESTED FACTORIAL EXPERIMENT

It is often impossible or undesirable to do a complete factorial design in chemical work. For example, interlaboratory studies of precision furnish data collected by different analysts on different instruments, so that we have Analysts a, b, and c using Instruments 1, 2, and 3 in one laboratory, studying the same methods as Analysts d, e, and f using Instruments 4, 5, and 6 in another laboratory. These conditions give rise to a series of small factorial designs nested within each other.

Example 3.6

In this example, we are comparing the effectiveness of two drying ovens and of two trays in each oven. Obviously, Trays 1 and 2 in Oven A are different from Trays 3 and 4 in Oven B. We are taking a sample from the front and back of each tray and running replicate analyses on each sample so taken (see Table 3.19).

Table 3.19

	Oven A				Oven B			
	Tray 1 Position		*Tray 2* Position		*Tray 3* Position		*Tray 4* Position	
Analysis	F	B	F	B	F	B	F	B
1	11.2	11.3	11.5	11.4	11.9	11.7	12.0	11.8
2	11.1	11.2	11.5	11.5	11.6	11.8	11.9	12.1
Sum position	22.3	22.5	23.0	22.9	23.5	23.5	23.9	23.9
Sum tray	44.8		45.9		47.0		47.8	
Sum oven	90.7				94.8			
Grand sum	185.5							

The model for this experiment is

$$X_{ijkl} = \mu + O_i + T_{j(i)} + P_{k(ij)} + E_{l(ijk)}.$$

The analysis of this experiment proceeds as follows:

Step 1. Correction is $(185.5)^2/16 = 2150.6406$.

Step 2. Total SS: $(11.2)^2 + (11.1)^2 + \cdots + (12.1)^2 = 2152.05$.

Step 3. SS of positions in trays:
$$[(22.3)^2 + (22.5)^2 + \cdots + (23.9)^2]/2 = 2151.9350.$$

Step 4. SS of trays in ovens:
$$[(44.8)^2 + (45.9)^2 + (47.0)^2 + (47.8)^2]/4 = 2151.9225.$$

Step 5. SS of ovens:
$$[(90.7)^2 + (94.8)^2]/8 = 1251.6912.$$

Step 6. Set up the ANOVA table (Table 3.20).

Table 3.20

Source	df	Sum squares
Between ovens	$(o - 1)$	SS ovens–SS correction (Step 5–Step 1)
Between trays in ovens	$o(t - 1)$	SS trays–SS ovens (Step 4–Step 5)
Between positions in trays in ovens	$ot(p-1)$	SS positions–SS trays (Step 3–Step 4)
Between analyses in positions in trays in ovens	$opt(a - 1)$	Total SS–SS positions (Step 2–Step 3)
Totals	$opta - 1$	(Step 2–Step 1)

Placing the figures in the ANOVA table we have

Table 3.21

Source	*df*	Sum squares	Mean square
Between ovens	1	1.0506	1.0506
Between trays in ovens	2	0.2313	0.1156
Between positions in trays in ovens	4	0.0125	0.0031
Between analyses in positions in trays in ovens	8	0.1150	0.0144

The ANOVA table for a nested factorial is slightly different from those previously encountered. Each source of variance is "nested" within the next larger one; e.g., "trays" are nested within "ovens" so that each source can be regarded as an individual analysis of variance.

The sources of variance are listed in the column labeled "source." Note the designations: "between trays within ovens," "between positions in trays in ovens," etc., instead of simply "between trays" and "between positions."

If we designate the number of ovens by o, trays by t, positions by p, and analyses by a, we can calculate the number of degrees of freedom for each position. Since ovens are not "within" anything, the degrees of freedom for ovens is $o - 1$. However, because t trays are within o ovens, degrees of freedom for "trays within ovens" is the degrees of freedom for trays $(t - 1)$ multiplied by the number of ovens (o). The *df* column gives the symbolic method of calculating degrees of freedom for each term.

The sum of squares column is composed of the sum of squares for the source corrected by subtracting the sum of squares of the source in which it is nested.

Step 7.   Components of variance

This is a mixed model. If there are only two trays and only two ovens, tray and oven factors are fixed, but the samples taken at the front and back of the trays must be random, making the position factor random.

Let us designate the number of ovens as $o = 2$, the number of trays in ovens as $t = 2$, the number of positions in trays in ovens as $p = 2$, and the number of analyses in positions in trays in ovens as $a = 2$. Then the mean square is an estimate of

Ovens: $V_e + taV_p + aptV_o = 1.0506;$

Trays: $V_e + taV_p + apV_t = 0.1156;$

Position: $V_e + taV_p = 0.0031;$

Error: $V_e = 0.0144.$

When testing for significance, position is tested against error, but trays and ovens are tested against positions, for the reasons given in Section 3.8.

If we have more than two trays in each oven, sampling from only two trays makes the tray factor a random variable. However, since we have only two ovens we still have a mixed model, and the mean square is an estimate of

Ovens: $V_e + taV_p + apV_t + aptV_o = 1.0506;$

Trays: $V_e + taV_p + apV_t = 0.1156;$

Positions $V_e + taV_p = 0.0031;$

Error: $V_e = 0.0144.$

When testing for significance, positions is tested against error, trays is tested against positions, and ovens is tested against trays.

Tests of Significance: "Positions in trays" obviously is not significant. It can be combined with V_e to give a new estimate of error.

$$(0.0125 + 0.1150)/(4 + 8) = 0.0106.$$

Mixed model (*trays fixed*)
  Trays:
$F = (0.1156)/0.0106 = 10.9,$
$F_{0.01}$ for $f_1 = 2$ and $f_2 = 12$ is 6.93.
  Ovens:
$F = 1.0506/0.0106 = 99.1,$
$F_{0.01}$ for $f_1 = 1$ and $f_2 = 12$ is 9.33.

Random model (*trays random*)
  Ovens:
$F = 1.0506/0.1156 = 9.09,$
$F_{0.05}$ for $f_1 = 1$ and $f_2 = 2$ is 18.5.

4

The Comparison of Two Averages

4.1 THE *t* TEST

As we have seen in the previous chapter, the *F* test is based on the assumption that there is no significant difference between variances, and therefore any significant difference must be caused by the differences in averages. When we are comparing two averages we make use of the *t* test, which is a special case of the *F* test.

The value for *t* is the square root of *F* at

$$f_1 = n - 1, \qquad f_2 = 1,$$

where *n* is the number of observations comprising the average.

Because it is a special case of the *F* test, the basic assumptions regarding the *F* test also apply to the *t* test. The following equation is the general equation for *t*.

$$t = (\bar{X}_1 - \bar{X}_2)/s_{\bar{X}}. \tag{4.1}$$

The distribution of *t* was mentioned in Section 1.3, and a curve of the distribution was shown in Fig. 1.1, curve b. The curve is symmetrical about $\bar{X}$ with two tails. When running tests of significance, we may use one or both tails of the curve. If we wish to know if $\bar{X}_1$ is greater or less than $\bar{X}_2$, we use only one tail of the *t* curve. This is called a one-tailed test. On the other hand, if we wish to know if $\bar{X}_1$ is part of the population of which $\bar{X}_2$ is the average, we make use of the whole *t* distribution, and we use a two-tailed test.

4.2 USES OF A t TEST

4.2.1 *Comparison of an Average with a Standard*

Example 4.1

The following results were obtained from an analysis: 100.3%, 99.2%, 99.4%, 100.0%, 99.4%, 99.9%, 99.4%, 100.1%, 99.4%, 99.6%.

The question is, within the limits of experimental error, can we say the results indicate 100% purity?

$$H_0: \quad \bar{X} = \mu = 100;$$
$$H_1: \quad \bar{X} < \mu = 100.$$
$$s_{\bar{X}} = s/\sqrt{n}, \quad \bar{X}_1 = \mu - 100,$$

hence Eq. (4.1) becomes

$$t = (\mu - \bar{X})\sqrt{n}/s. \tag{4.2}$$

Step 1. Calculate

$$\bar{X} = (100.3 + 99.2 + \cdots + 99.6)/10 = 997/10 = 99.7.$$

Step 2. Calculate s

$$s = \sqrt{\frac{X^2 - (\Sigma X)^2/n}{n - 1}},$$

$$s = \sqrt{\frac{(100.3)^2 + (99.2)^2 + \cdots + (99.6)^2 + (99.7)^2/10}{9}},$$

$$s = 0.362.$$

Step 3. Substitute in Eq. (4.2)

$$t = \frac{(100 - 99.7)\sqrt{10}}{0.362},$$

$$t = 2.62 \quad (f = n - 1 = 9).$$

In Table XI, the value for one-tailed $t_{0.05}$ is 1.83 for 9 *df*. The experimental value exceeds this, and therefore the null hypothesis is rejected. This analysis cannot be considered as indicating the material is 100% pure.

Example 4.2

If the question is: "Do these data come from a population whose average is 100%?", the test is a two-tailed test.

$$H_0: \quad \bar{X} = \mu = 100;$$

$$H_1: \quad \bar{X}_1 \neq \mu.$$

The calculation of t is the same as in Example 4.1,

$$t = 2.62.$$

Step 1.   Consulting Table XI, we find $t_{0.05} = 2.26$ for 9 *df* for a two-tailed test and we again reject the null hypothesis.

4.2.2 *Unpaired Replication*

The t test can be used to compare the difference between two averages both of which have a degree of uncertainty. We calculate $s_{\bar{X}}$ from the pooled variances of the two averages and Eq. (4.1) becomes

$$t = \frac{\bar{X}_1 - \bar{X}_2}{s} \sqrt{\frac{n_1 \times n_2}{n_1 + n_2}}. \tag{4.3}$$

Note that when $n_1 = n_2$ the expression under the radical reduces to $n/2$ and Eq. (4.3) becomes

$$t = \frac{\bar{X}_1 - \bar{X}_2}{s} \sqrt{\frac{n}{2}}. \tag{4.4}$$

Example 4.3

The following values were obtained by two analysts; is the difference between them significant? That is to say, can their results be considered as coming from the same population?

	Analyst 1	Analyst 2
	93.08	93.95
	91.36	93.42
	91.60	92.20
	91.91	92.46
	92.79	92.73
	92.80	94.31
	91.03	92.94
	—	93.66
	—	92.05
Average	92.08	93.08

$$H_0: \quad \bar{X}_1 = \bar{X}_2; \qquad H_1: \quad \bar{X}_1 \neq \bar{X}_2.$$

The calculations are similar to those in Example 4.1 except that the variances are pooled.

$$s_1{}^2 = (93.08)^2 + (91.36)^2 + \cdots + (91.03)^2 - (644.67)^2/7$$
$$= 3.9027;$$

$$s_2{}^2 = (93.95)^2 + (93.42)^2 + \cdots + (92.05)^2 - (837.72)^2/9$$
$$= 5.0836;$$

$$s = \left(\frac{s_1{}^2 + s_2{}^2}{(n_1 - 1) + (n_2 - 1)} \right)^{1/2}$$
$$= \left(\frac{3.9027 + 5.0836}{6 + 8} \right)^{1/2}$$
$$= 0.80.$$

Substitute in Eq. (4.3)

$$t = (93.08 - 92.08)/0.80\sqrt{(7 \times 9)/(7 + 9)},$$

$$t = 2.48 \qquad (f = n_1 - 1 + n_2 - 1) = 14.$$

From Table XI the critical value for 14 *df* is 2.145 and the null hypothesis is rejected.

4.2.3 *Paired Comparisons*

There are many occasions when experiments are done in pairs for the purpose of making comparisons. A few such occasions are

1. Comparison of two replications of several analyses,
2. Comparison of two methods of analysis by several different analysts,
3. Comparison of input and output figures of a reaction,
4. Comparison of the effects of a drug on experimental animals.

The reader can add many more from experience.

In all of these comparisons the object of the experiment is to learn if, on the average, there is a difference between the results. To do this, we compare the average difference between pairs with the variability between samples.

In this case, the standard deviation we use for comparison is the standard deviation of the difference between samples and $s_{\bar{X}}$ in Eq. (4.1) becomes s_d, changing Eq. (4.1) to

$$t = \frac{\bar{X}_1 - \bar{X}_2\sqrt{n}}{s_d}, \qquad (4.5)$$

where n is the number of differences.

Example 4.4 *Paired Replication*

Two laboratories analyze nine batches of phlogiston with the following results:

Batch	Lab A	Lab B	A – B
1	93.08	92.97	0.11
2	92.59	92.85	−0.26
3	91.36	91.86	−0.50
4	91.60	92.17	−0.57
5	91.91	92.33	−0.42
6	93.49	93.28	−0.21
7	92.03	92.30	−0.27
8	92.80	92.70	0.10
9	91.03	91.50	−0.47
$\bar{X}$	92.21	92.44	$d = -0.23$

$$H_0 = \bar{A} = \bar{B} \, ;$$
$$H_1 = \bar{A} \neq \bar{B} \, .$$

Step 1. Calculate the differences between Laboratory A and Laboratory B.

Step 2. $d = \text{Sum}(A - B)/n$
$= [0.11 + (-0.26) + \cdots + (-0.47)]/9$
$= -2.07/9$
$= -0.23.$

Step 3. $s_d{}^2 = [(0.11)^2 + (-0.26)^2 + \cdots + (-0.47)^2$
$- (-2.07)^2/9]/8 = 0.08785.$

Step 4. $s_d = (s_d{}^2)^{1/2}$
$= 0.296.$

Step 5.
$$t = \frac{(\bar{A} - \bar{B})\sqrt{n}}{s_d}$$

$$= \frac{(92.21 - 92.44)\sqrt{9}}{0.296}$$

$$= 2.33 \quad (df = n - 1 = 8).$$

The critical value is $t_{0.05} = 2.31$ for 8 *df*. The difference between the two laboratories is significant at the $(1 - \alpha) = 95\%$ level.

4.3 SUBSTITUTE *t* TESTS

It has been demonstrated by Lord (*3, 4*) that range may be substituted for standard deviation in Eq. (4.1) with only a small loss in efficiency. Equation (4.1) then becomes:

$$L = (\bar{X}_1 - \bar{X}_2)/R. \tag{4.6}$$

The critical values for L are given in Table III. For a one-tailed test, use the probabilities (0.05, 0.025, etc.) for the row P_1. For a two-tailed test the probabilities are in the row P_2. The column headed by n represents the number of observations. For example, to find the 0.05 critical value for a one-tailed test for five observations, use $P_1 = 0.05$ column, $n = 5$ row, and find the value 0.39.

When we have two averages, both of which have a degree of uncertainty, we can make use of the pooled variances of the averages to estimate t, or of the total range to estimate a substitute for t (5):

$$M = (\bar{X} - \bar{Y})/(R_x + R_y). \tag{4.7}$$

The critical values for M are given in Table IV. As in Table III, P_1 and P_2 refer to the probabilities associated with one-tailed and two-tailed tests, n_1 and n_2 are the number of observations in each of the two ranges.

For example, to find the 0.01 critical value for a two-tailed test with $n_1 = 2$, $n_2 = 5$, use $P_2 = 0.01$ column. Since $n_1 = 2$, use the $n_2 = 5$ row in the first group of numbers, and find the value 1.008.

4.4 USES OF SUBSTITUTE t TESTS

We will illustrate the use of the range tests by using the same examples as we used to illustrate the t test.

Example 4.5 *One-Tailed Test*

Using the data from Example 4.1:
100.3%, 99.2%, 99.4%, 100.0%, 99.7%, 99.9%, 99.4%, 100.1%, 99.4%, 99.6%.

$$H_0: \quad \bar{X} = \mu = 100;$$

$$H_1: \quad \bar{X} < \mu = 100$$

Step 1. $\bar{X} = 997/10 = 99.7$

Step 2. $R = 100.3 - 99.2 = 1.1$

Step 3. $L = (\mu - \bar{X})/R = (100 - 99.7)/1.1 = 0.272$

Step 4.   The critical value for a one-tailed test in Table III for $n = 10$ is $L_{0.05} = 0.19$. The calculated value exceeds this, and therefore H_0 is rejected.

Example 4.6 *Two-Tailed Test*

If we wish to know if these data came from a population whose average is 100%

$$H_0: \quad \bar{X} = \mu = 100;$$

$$H_1: \quad \bar{X} \neq \mu.$$

The value for L is calculated as above:

$$L = 0.27.$$

In Table III the critical value for $L_{0.05} = 0.23$ for a two-tailed test at $n = 10$ and the null hypothesis is again rejected.

Example 4.7 *Unpaired Replication Using Range*

Using the data in Example 4.3, we calculate:

Step 1. $\bar{X}_1 = 92.08, \qquad \bar{X}_2 = 93.08.$

Step 2. $R_1 = 93.08 - 91.03 = 2.05,$
$R_2 = 94.31 - 92.05 = 2.26.$

Step 3. Substitute in
$$M = (\bar{X}_2 - \bar{X}_1)/(R_2 + R_1)$$
$$= (93.08 - 92.08)/(2.26 + 2.05)$$
$$= 0.232.$$

Step 4. In Table IV the critical value, $M_{0.05} = 0.189$ for $n_1 = 7$, $n_2 = 9$ (and a two-tailed test). Hence, H_0 is rejected.

Example 4.8 *Paired Replication Using Range*

Example 4.4 can be done by range:

Batch	Lab A	Lab B	A – B
1	93.08	92.97	0.11
2	92.59	92.85	−0.26
3	91.36	91.86	−0.50
4	91.60	92.17	−0.57
5	91.91	92.33	−0.42
6	93.49	93.28	0.21
7	92.03	92.30	−0.27
8	92.80	92.70	0.10
9	91.03	91.50	−0.47
$\bar{X}$	92.21	92.44	$\bar{d} = -0.23$

Step 1.   Calculate $d = A - B$ for each laboratory

Step 2.   Calculate R_d; the range of $(A - B)$
$$R_d = 0.21 - (-0.57) = 0.78.$$

Step 3. $L = (\bar{X}_2 - \bar{X}_1)/R_d$
$$= (92.44 - 92.21)/0.78$$
$$= 0.295.$$

Step 4.   In Table III, the two-tailed critical value for $L_{0.05} = 0.255$, and therefore the null hypothesis is rejected.

5

Analysis of Variance by Range

5.1 INTRODUCTION

In Chapter 3, we demonstrated how statistically designed experiments are used in the comparison of several factors at several different levels, and we showed how the analysis of variance is used to analyze the data from these experiments. In this chapter, we shall replace the analysis of variance technique with the range methods of analysis.

These range methods are based upon the work of David (*1*), Hartley (*2*), and Patnaik (*3*) who extended the work of Lord. The purpose of the methods is to substitute range for variance as a measure of variability. Tests of significance are performed by the *F* ratio test, using variances calculated from range, and by the studentized range test:

$$q = (\bar{X}_1 - \bar{X}_2)/s_{\bar{X}}, \tag{5.1}$$

here, $s_{\bar{X}}$ is an independent estimate of the standard deviation obtained from the range. Critical values for q are listed in Table VII.

When the standard deviation and variance are calculated from range, there is a loss in degrees of freedom, making it necessary to use "equivalent degrees of freedom." An equivalent degree of freedom is approximately 90% of the regular degree of freedom. Tables V and VI give the equivalent degrees of freedom (f) for various combinations of k and n. These tables also list the factors c_1 and c_2 for converting the range into an unbiased estimate of the square root of the variance; that is to say:

$$\bar{R}/c_1 = \sqrt{V} = s. \tag{5.2}$$

To illustrate the use of range in place of standard deviation we shall use the same examples in this chapter that were used in Chapter 3.

5.2 BLOCK DESIGN: ONE-WAY CLASSIFICATION

Example 5.1

Example 3.1 postulated that in an experiment to test methods of drying, the methods were not all equally efficient. The results of the experiment are shown in Table 5.1. The calculation is as follows:

Table 5.1   Percentage of Water

Analysis	Method				
	1	2	3	4	
a	1.9	1.7	1.6	1.1	
b	1.7	1.6	1.7	1.3	
c	2.0	1.5	1.9	1.6	
d	1.6	2.0	1.6	1.1	
e	1.9	1.6	1.4	1.2	
Average	1.82	1.68	1.64	1.26	1.60
Range	0.4	0.5	0.5	0.5	

Step 1. $R = 0.4 + 0.5 + 0.5 + 0.5 = 1.9,$

$\bar{R} = 1.9/4 = 0.475,$

where $\bar{R}$ is the measure of the within-methods variation.

Step 2. The between-treatments variation is

$\bar{X}_1 - \bar{X}_4 = 1.82 - 1.26 = 0.56.$

In Table 5.1, there are four groups of five numbers: $k = 4, n = 5$. From Table V, $f = 14.7, c_1 = 2.37$.

Substituting the values for $\bar{R}$ and c_1 in Eq. (5.2):

$s = 0.475/2.37 = 0.20$

To determine if there is a significant difference between treatments we substitute in Eq. (5.1):

$$q = \frac{(\bar{X}_1 - \bar{X}_4)}{s_{\bar{X}}}$$

$$= \frac{0.56}{0.20/\sqrt{5}}$$

$$= \frac{0.56\sqrt{5}}{0.20} = 6.26.$$

The critical values for q are given in Table VII. Table VIIA gives $q_{0.05}$ values, and Table VIIB gives $q_{0.01}$ values.

We are comparing four averages: $k = 4$ at $14.7f$. Under the column for $k = 4$ we find that at $f = 14$, $q_{0.01} = 5.32$, and for $f = 15$, $q_{0.01} = 5.25$. Interpolating to $f = 14.7$ for $k = 4$, $q_{0.01} = 5.27$. The difference between treatments is significant.

Step 3. The analysis tells us that there is a significant difference between treatments, but it does not tell us how the treatments differ from one another.

Solving Eq. (5.1) for $\bar{X}_1 - \bar{X}_2$, we obtain

$$\bar{X}_1 - \bar{X}_2 = qs/\sqrt{n} = w. \tag{5.3}$$

When comparing two averages the value for q is found in Tables VIIA and B under $k = 2, f = 14.7 : q_{0.05} = 3.02$, $q_{0.01} = 4.18$.

Substituting in Eq. (5.3):

$$w_{0.05} = 3.02 \times \frac{0.200}{\sqrt{5}} = 0.270;$$

$$w_{0.01} = 4.18 \times \frac{0.200}{\sqrt{5}} = 0.373.$$

Any averages which differ more than 0.37 are significantly different at the 0.01 significance level: Treatment 4 is significantly lower than the other treatments at this level.

Step 4. Components of variance

In Example 5.1 we are investigating the only four available treatments, so we are dealing with a fixed model. The components of variance are:

Between treatments, $V_e + 5V_t = $ EMS for treatments.

Within treatments, $V_e = $ EMS for error.

We have one range of four treatments: $k = 1, \quad n = 4$.

The EMS for treatments can be calculated from the treatments range and the c_1 factor from Table V: $c_1 = 2.24$.

$$s_t = \frac{0.56\sqrt{5}}{2.24} = 0.56,$$

$$V_t = (s_t)^2 = (0.56)^2 = 0.3136 = \text{EMS}$$
for treatments.

$$5V_t = 0.3136 - V_e$$
$$= 0.3136 - s^2$$
$$= 0.3136 - (0.2)^2,$$
$$V_t = (0.3136 - 0.0400)/5 = 0.05472.$$

5.3 BLOCK DESIGN: TWO-WAY CLASSIFICATION

In Section 3.6, we synthesized a problem and analyzed it by ANOVA. The same data are tabulated in Table 5.2.

Table 5.2

		Method		Analyst average
Analyst	1	2	3	
A	10.1	7.8	12.0	9.97
B	8.9	7.1	10.9	8.97
C	11.1	9.2	12.9	11.07
Method average:	10.03	8.03	11.93	10.00

Example 5.2 *Two-Way Classification by Range*

Step 1. Calculate the residual variation. This is the variation that cannot be accounted for by the variation for methods or for analysts. It is, therefore, a measure of the error.

The calculation is done by first subtracting the method average from the individual. This value is then subtracted from both the analyst average and the grand average. For example, the residual for Analyst A, Method 1 is:

$$X - \bar{X}_m = 10.1 - 10.03 = \quad 0.07,$$
$$\bar{X}_a - \bar{\bar{X}} = 9.97 - 10.00 = -0.03,$$
$$\text{Residual} = \quad 0.10.$$

This is shown stepwise in Tables 5.2.1 and 5.2.2.

Table 5.2.1

Analyst	$X - \bar{X}_m$			$\bar{X}_a - \bar{\bar{X}}$
	1	2	3	
A	0.07	−0.23	0.07	−0.03
B	−1.13	−0.93	−1.03	−1.03
C	1.07	1.17	0.97	1.07

The residuals are calculated by subtracting $(\bar{X}_a - \bar{\bar{X}})$ from the values in Table 5.2.1, giving Table 5.2.2:

Table 5.2.2

Analyst	1	2	3
A	0.10	−0.20	0.10
B	−0.10	0.10	0.00
C	0.00	0.10	−0.10
Range:	0.20	0.30	0.20

Step 2. $\bar{R} = (0.20 + 0.30 + 0.20)/3 = 0.233$, $k = 3$, $n = 3$:
$c_2 = 1.48$ (from Table VIA).
$s = 0.233/1.48 = 0.157$, the standard deviation of residual.

Step 3. Methods range: $11.93 - 8.03 = 3.90$.

Step 4. Analyst range: $11.07 - 8.97 = 2.10$.
We have three methods and three analysts, hence $k = 3, n = 3$.

Step 5. Set up analysis of range table (Table 5.2.3).

Table 5.2.3 Analysis of Range

Source	k	n	f	R	s
Between methods	3	3	—	3.90	—
Between analysts	3	3	—	2.10	—
Residual	3	3	3.7	0.233	0.157

Step 6. Tests of significance:

Between analysts, $q = 2.10 \quad \sqrt{3}/0.157 = 23.2$.

Between methods $q = 3.9 \quad \sqrt{3}/0.157 = 43.0$.

From Table VIIB: At $f = 3.7$ and $k = 3$, $q_{0.01} = 8.86$. There is a significant difference between analysts and methods.

5.4 INTERACTION

Example 3.3 was synthesized by postulating that Analyst A commits a systematic error of $+0.5$ with Method 3 and Analyst B commits a systematic error of -0.5 with Method 3, and gave the data shown in Table 5.3. The analysis of these data by range is shown by the following procedure.

Example 5.3

Step 1. Calculate the range within analysts. This is the best measure of the experimental error.

$$R = (10.1 - 9.8) + (9.0 - 8.9) + \cdots + (12.6 - 12.4)$$
$$= 0.3 + 0.1 + 0.2 + 0.1 + 0.1 + 0.1 + 0.1 + 0.2 + 0.2$$
$$= 1.4.$$
$$\bar{R} = 1.4/9 = 0.155.$$
$$s = \bar{R}/c_1 = 0.155/1.15 = 0.135.$$

(c_1 from Table V: $k = 9, n = 2$.)

Table 5.3

	Method 1 Analyst			Method 2 Analyst			Method 3 Analyst		
Replication	A	B	C	A	B	C	A	B	C
1	10.1	8.9	11.1	7.8	7.1	9.2	12.5	10.9	12.4
2	9.8	9.0	10.9	7.9	7.2	9.1	12.4	11.1	12.6
Anal. aver.	9.95	8.95	11.0	7.85	7.15	9.15	12.45	11.0	12.5
Method aver.	9.97			8.05			11.98		
Grand average	10.00								

Step 2. The range of methods (between methods variation) is $11.98 - 8.05 = 3.93$.

Step 3. The range of analysts (between analysts variation) is $10.88 - 9.03 = 1.85$.

Table 5.4   Methods–Analysts

	Method			Analyst average
Analyst	1	2	3	
A	9.95	7.85	12.45	10.08
B	8.95	7.15	11.00	9.03
C	11.00	9.15	12.50	10.88
Methods average	9.97	8.05	11.98	10.00

Step 4. Calculate the interaction by making a methods–analyst table (Table 5.4) and calculating the M × A residual.

$$\bar{R} = (0.25 + 0.50 + 0.75 + 0.67 + 0.12 + 0.58)/6.$$

Residual range = 0.478.

Subtracting $\bar{X}_m$ from each value we obtain Table 5.5.

Table 5.5 $X - \bar{X}_m$

	Method		
Analyst	1	2	3
A	−0.02	−0.20	0.47
B	−1.02	−0.90	−0.98
C	1.03	1.10	0.52

Subtracting the value of $\bar{X}_a - \bar{\bar{X}}$ from the above values gives the residual table (Table 5.6).

Table 5.6   *Residuals* $(\bar{X} - \bar{X}_m) - (\bar{X}_a - \bar{\bar{X}})$

	Method			Range
Analyst	1	2	3	analysts
A	−0.10	−0.28	0.39	0.67
B	−0.05	0.07	−0.01	0.12
C	0.15	0.22	−0.36	0.58
Range methods	0.25	0.50	0.75	

Step 5.   Set up analysis of range table (Table 5.7).

Table 5.7

Source	k	n	f	R	s
Between methods	3	3	—	3.93	—
Between analysts	3	3	—	1.85	—
M × A Interaction	3	3	—	0.478	—
Error (within analysts)	9	2	8.1	0.155	0.135

Step 6.   Tests of significance:

Interaction: $q = 0.5$ $\sqrt{9}/0.135 = 11.1$.

Analysts: $q = 1.85$ $\sqrt{9}/0.135 = 41.1$.

Methods: $q = 3.93$ $\sqrt{9}/0.135 = 87.3$.

In Table VIIB at $f = 8.1$, $k = 3$: $q_{0.01} = 5.65$. All differences are significant.

5.5 THE LATIN SQUARE DESIGN

Section 3.10 gave an example of the analysis of the Latin Square design by ANOVA. The range method is illustrated in this section. Table 5.8 contains the same data as Table 3.10.

Example 5.4

Step 1.   Calculate the error term.

    (a) Remove the variation due to T by subtracting $\bar{X}_t$ from each value in each column. For example, $1.1 - 1.73 = -0.63$. This gives the values shown in Table 5.9.1.

    (b) Remove variation due to P by subtracting $(\bar{X}_p - \bar{\bar{X}})$ from each value in each row of Table 5.9.1. For example, $(\bar{X}_p = 1.73, \bar{\bar{X}} = 2.20)$: $-0.63 - (1.73 - 2.20) = -0.16$. This gives Table 5.9.2.

Table 5.8

	T_1	T_2	T_3	$\bar{X}_p$
P_1	1.1	1.7	2.4	1.73
P_2	1.9	2.4	2.5	2.27
P_3	2.2	2.6	3.0	2.60
$\bar{X}_t$	1.73	2.23	2.63	
$\bar{X}_c$	2.07	2.20	2.33	
$\bar{X}$	2.20	—	—	

Table 5.9.1

	T_1	T_2	T_3
P_1	−0.63	−0.53	−0.23
P_2	0.17	0.17	−0.13
P_3	0.47	0.37	0.37

Table 5.9.2

	T_1	T_2	T_3
P_1	−0.16 (C_1)	−0.06 (C_2)	0.24 (C_3)
P_2	0.10 (C_2)	0.10 (C_3)	−0.20 (C_1)
P_3	0.07 (C_3)	−0.03 (C_1)	−0.03 (C_2)

(c) Remove variation due to C by subtracting $(\bar{X}_c - \bar{\bar{X}})$ from each value for each catalyst in Table 5.9.2. For example, $(\bar{X}_c = 2.07, \bar{\bar{X}} = 2.20)$: $-0.16 - (2.07 - 2.20) = -0.03$. This gives Table 5.9.3.

Table 5.9.3

	T_1	T_2	T_3	R_p
P_1	−0.03	−0.06	0.11	0.17
P_2	0.10	−0.03	−0.07	0.17
P_3	−0.06	0.10	−0.03	0.16
R_t	0.16	0.16	0.18	—

The range of both columns and rows is averaged to avoid the errors introduced by rounding off figures.

$\bar{R} = (0.50 + 0.50)/6 = 0.167$.

Use Table VIB. Under $n = 3$, $k = 3$, we find

$f' = 3.6$, $c_3 = 1.40$.

Each n is made up of $m = 3$ values.

$$f' = mf. \tag{5.3}$$

Substituting in Eq. (5.3) and solving for f:

$f = 3.6/3 = 1.2$.

An estimation of s is made by use of Eq. (5.2)

$s = 0.167/1.40 = 0.119$.

Step 2.   Calculate the ranges of the main effects:

$R_t = 2.63 - 1.73 = 0.90$;

$R_p = 2.60 - 1.73 = 0.87$;

$R_c = 2.33 - 2.07 = 0.26$.

Step 3. Set up analysis of range table (Table 5.10)

Table 5.10

Source	k	f	$\bar{R}$	s
Between temperatures	1	—	0.90	—
Between pressures	1	—	0.87	—
Between catalysts	1	—	0.26	—
Error	—	1.2	0.167	0.119

Step 4. Calculate components of variance. The experiment involves three levels of temperature, pressure, and catalyst, and the data are applicable only to the conditions of the experiment. All three components are fixed, giving us a fixed model.

For each factor we have one range of three averages of three individuals, so that $k = 1$, $n = 3$, and $c_1 = 1.91$. Making use of the rules of Section 3.7, we obtain

$$EMS(T) = 3(0.9/1.91)^2 = 0.6600 = V_e + 3V_t;$$
$$EMS(P) = 3(0.87/1.91)^2 = 0.6173 = V_e + 3V_p.$$
$$EMS(C) = 3(0.26/1.91)^2 = 0.0555 = V_e + 3V_c.$$
$$\text{Error} = (0.119)^2 = 0.0142 = V_e.$$

The tests of significance are all based upon the ratios of the main effects and the error effect.

Step 5. Tests of significance. Substitute in Eq. (5.1):

Temperature, $\quad q = 0.90 \quad \sqrt{3}/0.119 = 13.1.$
Pressure, $\quad q = 0.87 \quad \sqrt{3}/0.119 = 12.6.$
Catalyst, $\quad q = 0.26 \quad \sqrt{3}/0.119 = 3.78.$

Table VII gives the critical value of q for $f = 1.2$, $k = 3$: $q_{0.05} = 23.0.$

Step 6.   Interpretation of results. There is no statistical evidence that the apparent effects of temperature, pressure, and catalyst are significant. The probability is greater than 5 chances in 100 that the differences are within the limits of experimental error. Efforts should be made to reduce this error, or the levels of the factors should be spaced farther apart.

5.6 FACTORIAL EXPERIMENTS

Example 5.5 *Two Factors, One with Replication*

In Section 3.4, we performed an analysis of variance of a two-factor experiment, one factor of which was replicated (Example 3.2). We shall now analyze this experiment by the range method.

Step 1.   Calculate the range of the replicate analyses for each analyst:

$$10.1 - 9.8 = 0.3,$$
$$9.0 - 8.9 = 0.1,$$
$$11.1 - 10.9 = 0.2,$$
$$7.9 - 7.8 = 0.1,$$
$$7.2 - 7.1 = 0.1,$$
$$9.2 - 9.1 = 0.1,$$
$$12.5 - 12.4 = 0.1,$$
$$11.1 - 10.9 = 0.2,$$
$$12.6 - 12.4 = 0.2.$$
$$R = 1.4;$$
$$\bar{R} = 1.4/9 = 0.155.$$

In Table V at $k = 9$, $n = 2$: $f = 8.1$, $c_1 = 1.15$. Substituting in Eq. (5.2):

$$s = 0.155/1.15 = 0.13;$$
$$V = 0.0169.$$

Step 2. Calculate the M × A interaction. We form the M × A Table (see Table 5.11) and calculate the average for each method ($\bar{X}_m$), the average for analysts ($\bar{X}_a$), and ($\bar{X}_a - \bar{\bar{X}}$).

Table 5.11

Analyst	Method 1	2	3	Sum	$\bar{X}_a$	$(\bar{X}_a - \bar{\bar{X}})$
a	19.9	15.7	24.9	60.5	20.2	0.2
b	17.9	14.3	22.0	54.2	18.0	−2.0
c	22.0	18.3	25.0	65.3	21.8	1.8
Sum	59.8	48.3	71.9	180.0		
$\bar{X}_m$	19.9	16.1	24.0	20.0		

Step 3. Subtract each $\bar{X}_m$ from individual values in Table 5.11; e.g., $19.9 - 19.9 = 0$. $17.9 - 19.9 = -2.0$. $22.0 - 19.9 = 2.1$. Tabulate these in Table 5.12.1.

Table 5.12.1

Analyst	Method 1	2	3
a	0.0	−0.4	0.9
b	−2.0	−1.8	−2.0
c	2.1	2.2	1.0

Step 4. Subtract ($\bar{X}_a - \bar{\bar{X}}$) from the values in the corresponding rows of Table 5.12.1 and calculate the ranges of the columns and rows (see Table 5.12.2).

Table 5.12.2

Analyst	Method 1	Method 2	Method 3	R_a
a	−0.2	−0.6	0.7	1.3
b	0	0.2	0	0.2
c	0.3	0.4	−0.8	1.2
R_m	0.5	1.0	1.5	2.7
		$R = 3.0$		

Except for the effect of rounding off figures, sums of columns and rows should equal zero.

$\bar{R}_{ma} = (3.0 + 2.7)/[(3 \times 2) + (3 \times 2)] = 0.48.$

In Table VI, at $k = 3$, $n = 3$: $f = 3.7$, $c_2 = 1.48$. Substituting in Eq. (5.1):

$s = 0.48\sqrt{2}/1.48 = 0.46;$

$V_{ma} = 0.2116.$

Step 5. Calculate the range of each main effect:

Methods: $R_m = (71.9 - 48.3)/6 = 3.93.$

Analysts: $R_a = (65.3 - 54.2)/6 = 1.85.$

Step 6. Set up analysis of range table (Table 5.13).

Step 7. Determine the components of variance. At this point, we must determine whether this experiment should be described by a fixed or mixed model. It would be fixed if we were interested only in the results of these particular three analysts with these three methods. It would be mixed if we wished to project our results to cover the expected variation

Table 5.13

Source	k	f	R	s	V
Between methods	3	—	3.93	—	—
Between analysts	2	—	1.85	—	—
M × A	3	3.7	0.48	0.46	0.2116
Within analysts	—	8.1	0.155	0.13	0.0169

between analysts in general, with these particular methods of analysis. It could not be a random model because we could not project our results to cover methods of analysis in general. The choice of models rests with the experimenter; we shall determine the components of variance for both models.

For both methods and analysts we have one range of three averages of six individuals: $k = 1$, $n = 3$, $c_1 = 1.91$. Following the rules in Section 3.7, we can set up Table 5.14.

Table 5.14   Components of Variance

Source	EMS	Fixed model	Mixed model
Methods	$6(3.93/1.91)^2 = 25.40$	$V_e + 6V_m$	$V_e + 2V_{ma} + 6V_m$
Analysts	$6(1.85/1.91)^2 = 5.64$	$V_e + 6V_a$	$V_e + 6V_a$
M × A	0.2116	$V_e + 2V_{ma}$	$V_e + 2V_{ma}$
Error	0.0169	V_e	V_e

Solving the equations in Table 5.14, we obtain the following individual variances:

	Fixed model	Mixed model
V_m	3.8771	3.8430
V_a	0.6247	0.6247
V_{ma}	0.1024	0.1024
V_e	0.0169	0.0169

These values are in excellent agreement with those found by the analysis of variance.

Step 8. Run tests of significance.

(1) M × A Interaction

$$F = 0.2116/0.0169 = 12.52.$$

In Table VIIIA, $F_{0.05} = 4.0$ and $F_{0.01} = 7.1$ for $f_1 = 3.7, f_2 = 8.1$.

(2) Variation between analysts

$$q = \frac{1.85\sqrt{6}}{0.13} = 3.78.$$

(3) Variation between methods (fixed model)

$$q = \frac{3.93\sqrt{6}}{0.13} = 74.0.$$

In Table VIIA, at $k = 2$, $f = 8.1$: $q_{0.05} = 3.26$ and $q_{0.01} = 4.74$.

(4) Variation between methods (mixed model).

In order to test for significance for methods as a mixed model we test methods against methods–analysts:

$$q = \frac{3.93\sqrt{6}}{0.46} = 20.9.$$

In Table VIIA at $k = 3, f = 3.7$: $q_{0.05} = 5.26$.

Step 9.   Interpretation of tests of significance.

There is an interaction between methods and analysts. This means that at least one analyst is making a systematic error with one of the methods. If the reader refers to Example 3.2, he will remember that this problem was set up with the hypothesis that Analysts A and B were making a systematic error with Method 3.

There is a difference between methods as a fixed model, but not as a mixed model, because the methods–analysts interaction has covered up the significant difference. If the analysts were not committing a systematic error, the significance of the difference in methods would be apparent.

There is also a difference between analysts.

Although the primary reason for the experiment was to determine if there were a difference between methods, the experiment was also designed to detect differences between analysts and the methods–analysts interaction. The fact that these factors were significant is important information. Obviously, before any work can be done to correct differences in methods, the analysts' techniques must be scrutinized with a view to correcting the reasons for (a) the difference between their results, and (b) the systematic error which is causing the interaction.

Example 5.6 *Nested Factorial by Range*

The range method can also be used to analyze a nested factorial. Using the data from Example 3.6:

Step 1.   Calculate range and variance between ovens:

$R = (94.8 - 90.7)/8 = 0.5125.$

There is one range of two averages, each average being made up of eight observations. Therefore, $m = 8$. From Table V at $k = 1, n = 2$: $f = 1.0, c_1 = 1.41$:

$V = 8(0.5125/1.41)^2 = 1.0568.$

Step 2.   Calculate range and variance for trays in ovens:

Oven A $(45.9 - 44.8)/4 = 0.275$

Oven B $(47.8 - 47.0)/4 = \underline{0.200}$

Sum $R = 0.475$

$\bar{R} = 0.475/2 = 0.2375$

There are two trays in each oven. Each tray average is the average of four individual analyses; therefore, $m = 4$. From Table V at $k = 2, n = 2$: $f = 1.9$, $c_1 = 1.28$:

$V = 4(0.2375/1.28)^2 = 0.1376$.

Step 3.   Calculate range and variance for positions in tray:

Tray 1 $(22.5 - 22.3)/2 = 0.10$

Tray 2 $(23.0 - 22.9)/2 = 0.05$

Tray 3 $(23.5 - 23.5)/2 = 0.00$

Tray 4 $(23.9 - 23.9)/2 = \underline{0.00}$

Sum $R = 0.15$

$\bar{R} = 0.15/4 = 0.0375$.

There are four ranges in each position; each position average is the average of two individual analyses; therefore, $m = 2$. From Table V at $k = 4$, $n = 2$: $f = 3.7$, $c_1 = 1.21$:

$V = 2(0.0375/1.21)^2 = 0.0019$.

Step 4.   Calculate range and variance between analyses:

$11.2 - 11.1 = 0.1$,

$11.3 - 11.2 = 0.1$,   etc.

Sum $R = 1.1$,

$\bar{R} = 1.1/8 = 0.1375$.

In Table V, at $k = 8, n = 2$: $f = 7.2, c_1 = 11.6$:

$s = 0.1375/1.16 = 0.118$,

$V = 0.0139$.

Step 5. Set up analysis of range table

Analysis of Range

Source	n	f	R	V
Between ovens	2	1.0	0.5125	1.0568
Between trays in ovens	2	1.9	0.2375	0.1376
Between positions in trays in ovens	2	3.7	0.0375	0.0019
Between analyses	2	7.2	0.1375	0.0139
	$s = 0.118$			

"Positions in trays" obviously is not significant. It can be combined with V_e to give a new estimate of error as shown in the next tabulation.

$$f \times V = MS$$

$$7.2 \times 0.0139 = 0.1000$$
$$3.7 \times 0.0019 = 0.0070$$
$$\overline{10.9} \qquad \overline{0.1070}$$

$$V_e^1 = 0.1070/10.9 = 0.0098$$
$$s_e = 0.1$$

Step 6. Components of variance. This is a mixed model (see Section 3.12, Example 3.6, Step 7).

The mean square is an estimate of

Ovens: $V_e + taV_p + aptV_0$.

Trays: $V_e + taV_p + apV_t$.

Position: $V_e + taV_p$.

Error: V_e.

When testing for significance position is tested against error but trays and ovens are tested against positions. If we have more than two trays in each oven, sampling from two trays makes the tray variable a random variable, and the mean square is an estimate of

Ovens: $V_e + taV_p + apV_t + aptV_o$.

Trays: $V_e + taV_p + apV_t$.

Positions: $V_e + taV_p$.

Error: V_e.

When testing for significance position is tested against error, trays is tested against positions, and ovens is tested against trays.

When trays are random

$V_e = 0.0098$,

$V_p = 0.0000$,

$V_t = 0.0309$,

$V_o = 0.1149$.

Step 7. Tests of significance

Trays in ovens: $q = 0.2375$ $\sqrt{4}/0.1 = 4.75$.

In Table VIIB at $n = 2, f = 10.9$: $q_{0.01} = 4.39$.

Between ovens:

    (a) Trays fixed: $q = 0.5125 \sqrt{8}/0.1 = 14.38$

    (b) Trays random: $F = 1.0568/0.1376 = 7.68$

In Table VIIIA at $f_1 = 1.0, f_2 = 1.9$: $F_{0.05} = 18.5$.

Step 8. Interpretation of results

The efficiency of the drying operation depends upon which trays in the ovens are being used. There is a difference between ovens if only these four trays are to be considered. However, if the trays are merely a sample of possible trays in the ovens, the difference is not significant.

REFERENCES

1. David, H. A. (1951). Further applications of range to analysis of variance. *Biometrika* **38,** 393.
2. Hartley, H. O. (1950). The use of range in Anova. *Biometrika* **37,** 271.
3. Patnaik, P. B. (1950). The use of mean range in statistical tests. *Biometrika* **37,** 78.
4. Wernimont, G. (1951). Design and interpretation of interlaboratory studies of test methods. *Anal. Chem.* **23,** 1572.

6

Control Charts

6.1 INTRODUCTION

The control chart is a technique invented by Dr. Walter A. Shewhart (2) of the Bell Telephone Laboratories. Although originally intended for controlling the level of a quality characteristic for manufacturing processes, it has been found to be a useful statistical tool with broad applications.

There are two basic types of control charts: (1) attribute charts, which chart the number or percentage of defects in a process, and (2) variables charts, which are charts of numerical measurements made on the quality characteristic. Variables charts are of more use to chemists and chemical engineers and are the type to be described in this chapter.

The variables control chart has been defined as "A graphic representation of test data in such a manner that the variability of all the results is compared with the average variability within arbitrary small groups of the test data (3)." It is, in effect, a graphic, one-way analysis of variance.

6.2 NOMENCLATURE

Throughout this chapter we shall use the nomenclature that has become standard for control charts in the United States. It differs slightly from the terminology used elsewhere in this book (see Table 6.1).

Table 6.1

Symbol[a]	Definition
X	A single observation
$\bar{X}$	An average of a subgroup of observations $(\Sigma X/n)$
$\overline{\overline{X}}$	The grand average $(\Sigma X/kn = \Sigma \bar{X}/k)$
n	The number of observations in a subgroup
k	The number of subgroups
R	The range of a subgroup; the difference between the largest and smallest value
$\bar{R}$	The average range $(\Sigma R/k)$
σ	The standard deviation (root mean square) of a set of n numbers from their average $[\Sigma(X - \bar{X})^2/n]^{1/2}$
$\bar{\sigma}$	Average standard deviation $(\Sigma \bar{\sigma}/k)$
$\sigma_{\bar{x}}$	Standard error of the mean (σ/n)

[a] Any of the symbols with a prime (e.g., $\bar{X}'$) represent the parameter of the universe.

6.3 THEORY OF CONTROL CHARTS

Shewhart postulated that the measured characteristic of a manufactured item is subject to a small amount of variation due to chance alone. The chance system of variation is stable unless acted upon by some outside cause. This chance variation is inherent in any system of production, inspection, measurement, or testing. It is possible to measure and to plot the chance variation and by means of control limits generated by the system itself, to detect the existence of an outside force when one acts upon the system.

6.3.1. *Normality*

It is assumed that the population from which the sample data are drawn is statistically normal. It is true that a normal population is an ideal situation which is not always met in practice. However, it is a mathematical description of a constant-cause system—one in which nothing but chance variation is operating. In such a system, there should be approximately an equal number of variations above and below the arithmetical mean, and small variations should be more frequent than large ones. The concept of normality can, therefore, be used to test the probability that a measurement is from a system that is not influenced by anything but chance variations; or in other words, a system in statistical control.

6.4 CONTROL LIMITS

Normal distributions can be described by two parameters. These are the arithmetic mean (here called the average $\bar{X}$) and the standard deviation (σ).

The average locates the center of the distribution, and the standard deviation describes the spread of the data. It is known that in a normal distribution 68% of the values will be within $\bar{X} \pm 1\sigma$, 95% within $\bar{X} \pm 2\sigma$, and 99.7% within $\bar{X} \pm 3\sigma$.

This reasoning is the basis for setting the control limits on both the average and range charts. When a value is within the designated control limits it is assumed that only the normal chance variations are present in the system, and it is said to be "in control." A value outside of the control limits indicates that there is something causing more than the normal chance variation, and the value is commonly referred to as being "out of control."

Control limits for both the average chart and the range chart are calculated from the average range.

6.5 THE CHART FOR AVERAGES

This is the chart that shows the variation between groups of data. It is made by breaking the information into rational subgroups and plotting the subgroup averages as points on a graph. A solid line representing the grand average and broken lines 2σ or 3σ on either side of the grand average are drawn. The broken lines represent the control limits for the subgroup averages.

6.6 THE CHART FOR RANGES (OR STANDARD DEVIATIONS)

A second chart showing the variation within each subgroup is drawn below the average chart. Variation can be measured as either the range or the standard deviation. A solid line is drawn to represent the average variation and broken lines are drawn to represent the calculated control limits.

6.7 SUBGROUPS

The composition of the subgroup determines the amount and kind of information that can be derived from a control chart. Since the average of the group and the variability within it are the bases for judging the overall variability, it is important that the group should be chosen in some rational manner. For this reason the literature on control charts often refers to the groups as "rational groups" or "rational subgroups."

For example, if it were desired to control the weight of tablets being made on a tablet press with eight heads, a sample of eight consecutive tablets would constitute a rational subgroup. The average weight and the range of the weights would give the desired information about the variability of the tablet press.

In general, subgroups should be chosen so that a minimum variation exists within the subgroup.

6.7.1 *Subgroup Variability*

The variability within the subgroup is assumed to be the inherent variability which is due solely to uncontrollable chance factors. These are the variations usually called, "noise," "experimental error," etc.

Variability can be evaluated by means of the standard deviation or the range. The standard deviation is the root mean square of the data and is the most efficient method of evaluating variability. The range is the difference between the largest and smallest measurements. It is nearly as efficient as the standard deviation for small groups of data and is obviously much easier to calculate. Range is almost universally used in control chart work. The group size may vary from two to ten. Range loses its accuracy if the group size is more than ten.

6.7.2 *Subgroup Averages*

A basic assumption of the control chart technique is that the data are from a single normal population. However, if the individual measurements do deviate from normality, the control chart is not seriously affected when averages are used. It can be demonstrated that the averages of subgroups of nonnormal individual values are sufficiently close to a normal distribution for practical purposes.

6.8 CALCULATION OF CONTROL LIMITS

Control limits are calculated from the average range $(\bar{R})$ by means of the factors given in Table IX. Since $\bar{R}$ is a measure of the inherent variability of the data, it follows that the process being measured sets its own control limits and that these are independent of specification limits.

Since we assume normality, 99.7 % control limits should be set at

$$\bar{X} \pm 3\sigma'/\sqrt{n},$$

where σ' is the population standard deviation and n the subgroup size. However, σ' is not known to us; we are estimating it from $\bar{R}$. It has been found that, in the long run, $\bar{R}/\sigma'$ equals a constant (d_2) that varies with the size of the subgroup. Hence,

$$3\sigma'/\sqrt{n} = 3\bar{R}/d_2, \qquad n = A_2{}^*\bar{R},$$

where

$$A_2{}^* = 3/d_2\sqrt{n}.$$

As in all statistical work, the more data one has the more reliable are the statistics derived from them. In setting up control limits it is well to have at least 25 subgroups of size four or five upon which to base calculations.

However, one is often limited by practical considerations as to the size of the subgroup and also as to the number of subgroups. This is particularly true when the data are acquired from chemical analyses. For this reason Tables IXA, IXB, and IXC give values for $A_2{}^*$, $D_3{}^*$, and $D_4{}^*$, respectively. These constants are based on the work of F. S. Hillier (*1*) and are valid for small numbers of subgroups.

6.9 SIGNIFICANCE OF CONTROL LIMITS

There are two types of error inherent in all sampling procedures: Type 1 is the risk of rejecting a good lot and Type 2 is the risk of accepting a bad lot.

In the United States it is customary to use "3σ" limits for control charts. This means that a point falling outside of the limits would come from an acceptable lot only 3 times in a thousand and hence the probability of committing a Type 1 error is only 0.003. In Great Britain "warning limits" are set at 2σ (probability 0.05) and control limits at 3.09σ (probability 0.001).

6.10 RUNS

These schemes protect against the possibility of making a Type-1 error, but the risk of committing a Type-2 error is somewhat greater. For example, if the control chart is set up with the usual 3σ limits, a shift of the average by as much as one standard deviation would produce out of control points only 2.5% of the time. If 2σ limits are used, 16% of the values would be out of control.

The best way to guard against Type-2 errors is to watch the chart for runs: an excessive number of points on either side of the average line. Table 6.2 gives the probability of obtaining runs of a given length.

Table 6.2

Number on one side of $\overline{X}$	Out of a total of	Probability
7	7	0.0078
10	11	0.0054
12	14	0.0056
14	17	0.0052
16	20	0.0046

Table 6.2 shows that the probability of obtaining seven consecutive values on one side of the average is only 0.0078 from a process in control. Hence, such a situation is a warning that the process average may have shifted.

6.11 MAKING A CONTROL CHART

Example 6.1

Let us suppose we are running four replications of an analysis at a time. Consider data in Table 6.3 as the deviations in tenths of a percent from the theoretical values for 25 such analyses. (Actually, for the purposes of illustration the numbers were drawn, four at a time, with replacement from a population of 200 with $\overline{\overline{X}}' = 0$ and $\sigma' = 1.715$.)

Table 6.3[a]

k	X_1	X_2	X_3	X_4	$\bar{X}$	R	$\bar{\bar{X}}$	$\bar{R}$
1	−2	0	−1	−1	−1.00	2		
2	2	−1	−1	−2	−0.50	4		
3	2	0	1	0	0.75	2		
4	2	−3	1	1	0.25	5		
5	2	−2	0	0	0.00	4		
							−0.10	3.40
6	2	−2	−4	−1	−1.25	6		
7	0	1	0	−3	−0.50	4		
8	1	−3	2	−1	−0.25	5		
9	−2	−2	0	−2	−1.50	2		
10	−1	1	0	−1	−0.25	2		
							−0.425	3.60
11	−2	2	1	3	1.00	5		
12	2	0	−1	0	0.25	3		
13	2	−1	2	0	0.75	3		
14	0	−2	−1	2	−0.25	4		
15	1	0	0	0	0.25	1		
							−0.15	3.47

[a] Deviations in 0.1%.

Table 6.3ª (cont.)

k	X_1	X_2	X_3	X_4	$\bar{X}$	R	$\bar{\bar{X}}$	$\bar{R}$
16	0	1	−1	−1	−0.25	2		
17	1	0	−2	0	−0.25	3		
18	1	1	−1	−1	0.00	2		
19	0	2	0	−2	0.00	4		
20	2	−3	−3	−2	−1.50	5		
							−0.21	3.40
21	1	1	−3	−1	−0.50	4		
22	0	1	1	−1	0.25	2		
23	0	1	1	1	0.75	1		
24	−1	0	−5	1	−1.25	6		
25	1	2	1	1	1.25	1		
						$\bar{X}$	−0.15	
						$\bar{R}$		3.28

ª Deviations in 0.1%.

The calculations are as follows:

Step 1. Calculate the subgroup averages: $\bar{X} = \Sigma X/n$. For Subgroup 1, $\bar{X} = (-2) + 0 + (-1) + (-1) = -4/4 = -1$.

Step 2. Calculate subgroup ranges: $R = $ Largest $X - $ smallest X. For Subgroup 1, $R = 0 - (-2) = 2$.

Step 3. Calculate the grand average of the first five subgroups: $\bar{\bar{X}} = \Sigma \bar{X}/5 = -0.50/5 = -0.10$.

Step 4.   Calculate the average range of the first five subgroups: $\bar{R} = \Sigma R/5 = 17/5 = 3.40$.

Step 5.   Calculate the control limits.
(a) For averages: Using $A_2{}^*$ from Table IXA

$$CL = \bar{\bar{X}} \pm A_2{}^*\bar{R}$$
$$= (-0.10) \pm 0.952\,(3.4)$$
$$= -3.54 \text{ to } 3.34.$$

(b) For ranges: Using $D_3{}^*$ and $D_4{}^*$ from Tables IXB and IXC

$$CL = D_3{}^*\bar{R} \quad \text{and} \quad D_4{}^*\bar{R}.$$
$$D_3{}^*\bar{R} = 0.160(3.40) = 0.54;$$
$$D_4{}^*\bar{R} = 2.79(3.40) = 9.49.$$

Step 6.   Plot the points, drawing the average line, the average range, and the control limits for the five points. All points should be in control (see Fig. 6.1).

Step 7.   Repeat Steps 1–6 for every five subgroups until 25 subgroups have been collected:

Points	$\bar{\bar{X}}$	$\bar{R}$	$\bar{\bar{X}} \pm A_2{}^*\bar{R}$	$D_3{}^*\bar{R}$	$D_4{}^*R$
6–10	-0.42	3.60	-3.43–2.59	0.58	9.11
11–15	-0.15	3.47	-2.94–2.64	0.57	8.46
16–20	-0.21	3.40	-2.88–2.46	0.56	8.13
21–25	-0.15	3.28	-2.69–2.39	0.54	7.74

If no points are out of control, establish standard values for averages and control limits using the factors for infinity.

If, during the set-up period, there are points out of control, these values should be excluded from the calculation of control limits.

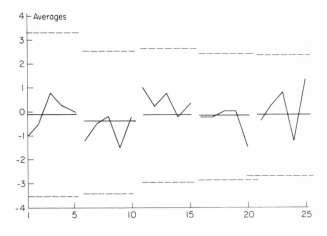

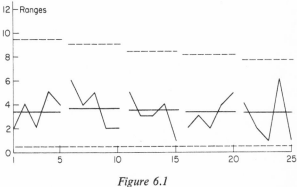

Figure 6.1

Figure 6.1 is a control chart that shows a perfect state of control. The analyst would be justified in claiming a precision of $\pm 0.25\%$ for the average of four assays by the method. He would expect that the highest and lowest analytical value from a group of four replications would not vary by more than 0.075%.

6.12 LACK OF CONTROL

Suppose an inexperienced analyst runs the method and obtains the results given in Table 6.4 and plotted in Fig. 6.2 which is an extension of Fig. 6.1. The control limits are calculated from the

Table 6.4

k	X_1	X_2	X_3	X_4	$\bar{X}$	R
26	2	2	0	-1	0.75	3
27	0	1	0	-1	0.00	2
28	3	-1	0	0	-1.00	3
29	1	-1	-1	0	-0.25	2
30	-2	2	-1	-4	-1.25	6
31	-2	3	0	1	0.50	5
32	5	1	1	4	2.75	4
33	2	-1	1	-3	-0.25	5
34	1	4	0	-1	1.00	5
35	-3	-4	2	-4	-2.25	6
36	2	3	-3	-1	0.25	6
37	-3	1	3	-4	-0.75	7
38	2	1	-1	1	0.75	3
39	0	3	-1	-2	0.00	5
40	1	-1	-5	6	0.25	11

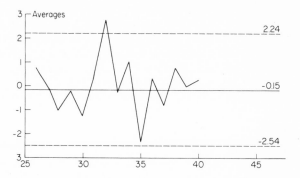

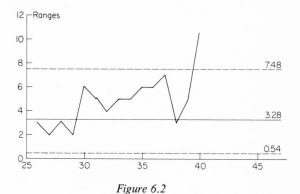

Figure 6.2

A_2^*, D_3^*, and D_4^* factors for infinity:

$$(-0.15) \pm 0.729(3.28) = -2.54\text{–}2.24;$$
$$0.166(3.28) = 0.54;$$
$$2.28(3.28) = 7.48.$$

It is obvious that the second analyst's results are less precise. The first indication can be found in Analysis 32, where the average is out of control. Analyses 30 through 37 inclusive give eight consecutive ranges above the average range, and the last analysis gives a range point which is out of control.

These data were generated by drawing four numbers with replacement from a normal population of 201 with $\bar{\bar{X}}' = 0$ and $\sigma' = 3.47$. It is interesting to note that even though the population average did not shift, one of the first indications of a change in the system was the high result in Analysis 32.

REFERENCES

1. Hillier, Frederick S. (1969). $\bar{X}$ and R chart control limits based on a small number of subgroups. *Quality Technology* **1,** 17.
2. Shewhart, W. A. (1931). "Economic Control of Quality of a Manufactured Product." Van Nostrand-Reinhold, Princeton, New Jersey.
3. Wernimont, G. (1946). Control charts in analytical laboratories. *Ind. Eng. Chem., Anal. Ed.* **18,** 587.

7

Correlated Variables

7.1 LINEAR REGRESSION

In scientific work, we often wish to determine the effect that some variable exerts on another. For example, We may wish to check a colorimetric reaction to see if it follows the Beer–Lambert Law, or to measure the rate of a chemical reaction, or to check the validity of a new assay method against a series of known standards.

A laboratory experiment is a measurement of the effect of one variable upon another: We measure a quantity of sample (X), carry out a reaction, and measure a response (Y). Expressed mathematically:

$$Y = f(X). \tag{7.1}$$

If the relationship is linear, Eq. (7.1) becomes:

$$Y = a + bX. \tag{7.2}$$

This is the equation of a straight line in which a is the intercept with the Y axis, and b is the slope of the line: the change in Y per unit change in X.

To a chemist a is a measure of a constant error, such as the blank of a titration or colorimetric analysis. The slope b may be the rate of a reaction or it may represent the ability of a test to distinguish between changes in X.

The study of the effect of one interrelated variable upon another is known as regression analysis. We are estimating the regression of Y upon X by use of Eq. (7.2), which is the best estimate of the line of regression.

In regression analysis we have two types of variables. The first, designated as independent variables, are those which can either be set to a desired level (for example, input temperature), or that can have values that can be observed but not controlled (i.e., ambient temperature). The dependent or response variables result from the changes that are made in the independent variable. This is a measurement which varies in a random fashion about its true value.

The regression line is fitted to the data by the method of least squares: The values of a and b in Eq. (7.2) are determined in such a way that the sum of squares of the Y values about the regression line are a minimum.

7.2 A LABORATORY USE OF REGRESSION

Let us suppose an analyst weighs a series of samples, $X_1, X_2, X_3, \ldots, X_n$, and carries out a reaction that gives results, $Y_1, Y_2, Y_3, \ldots, Y_n$. He may calculate the percentage of Y by the equation:

$$\% Y = 100 Y/X.$$

If the result contains a constant error (a), the analyst is measuring the ratio $100(Y - a)/X$, and not $100 Y/X$. If the weights are the same, he has no way of knowing if the bias (a) exists, and no way of measuring it; but if the values of $X_1, X_2, X_3, \ldots, X_n$ are significantly different he can use Eq. (7.2) to evaluate a.

Example 7.1

The standardization of sodium nitrite is accomplished by the titration of a weighed amount of Novocain in cold acid solution. The nitrous acid diazotizes the amino group of the Novocain. An excess of sodium nitrite forms a blue ring when touched to starch iodide paper. A blank of 0.1–0.2 ml is required to produce the end point.

Suppose an analyst performs the standardization without using the blank. He weighs four samples of 800 mg of Novocain and obtains an average titration of 29.60 ml. He would calculate the factor of the sodium nitrite as 0.9907 X 0.1 N. But this titration contains a bias of 0.15 ml due to the forgotten blank, so that the factor is really 0.9958 X 0.1 N, a relative error of approximately 5%.

Suppose the analyst weighs $X_1 = 200$ mg, $X_2 = 400$ mg, $X_3 = 600$ mg, and $X_4 = 800$ mg of Novocain and titrates these samples with the unstandardized sodium nitrite solution. His results are shown in the following tabulation:

Novocain (mg) (X)	Sodium nitrite (ml) (Y)
200	7.50
400	14.90
600	22.25
800	29.65
$\bar{X} = 500$	

Step 1. Calculate ΣX, ΣY, ΣXY, ΣX^2, and ΣY^2 (see Table 7.1). If we use $(X - \bar{X})$ instead of X in these calculations we will save a considerable amount of computation. This is equivalent to rewriting Eq. (7.2) as

$$Y = a + b(X - \bar{X}). \tag{7.3}$$

Step 2. Calculate the slope (b)
$$b = \Sigma(X - \bar{X})Y/\Sigma(X - \bar{X})^2; \tag{7.4}$$
$$b = 7380/200{,}000$$
$$= 0.03690.$$

Step 3. Calculate the intercept (a)
$$a = \bar{Y} - b\bar{X}; \tag{7.5}$$
$$a = 74.30/4 - 0.03690(5)$$
$$= 0.125.$$

Table 7.1

X (mg)	$X - \bar{X}$	Y (ml)	$(X - \bar{X})Y$	$(X - \bar{X})^2$
200.0	-300.0	7.50	-2250	90,000
400.0	-100.0	14.90	-1490	10,000
600.0	100.0	22.25	2225	10,000
800.0	300.0	29.65	8895	90,000

$$\bar{X} = 500$$

$$\Sigma X^2 = 1,200,000$$

$$\Sigma Y = 74.30$$

$$\Sigma(X - \bar{X})Y = 7380$$

$$\Sigma(X - \bar{X})^2 = 200,000$$

Step 4. Substitute these values in Eq. (7.2)

$$Y = 0.125 + 0.0369X. \tag{7.6}$$

Equation (7.6) is the best fitting relationship between milliliters of sodium nitrite and milligrams of Novocain under the conditions of the experiment.

Step 5. Calculate the factor of the normality of the sodium nitrite solution.

$$F = X/27.28\,Y, \tag{7.7}$$

where X is in milligrams of Novocain, and Y is in milliliters of sodium nitrite.

Hence, 1.0 ml 0.1 N $NaNO_2$ is equivalent to 27.28 mg of Novocain.

Step 5.1 Calculate the theoretical milliliters of sodium nitrite using Eq. (7.6). This gives the second column of Table 7.2.

Table 7.2

X (mg)	Y (ml)	F
200.0	7.505	0.9769
400.0	14.885	0.9851
600.0	22.265	0.9878
800.0	29.645	0.9892
	Average $F = 0.9848$	

Step 5.2 Calculate F, using Eq. (7.7) and the values for Y from Step 5.1.

The analyst, of course, wishes to know the precision of these results. In order to do this he must calculate the variance of Y with respect to X, and from this compute the standard deviations of the intercept (a) and the slope (b).

Example 7.2 *Calculation of $s_{y/x}$, $s(a)$, and $s(b)$.*

Step 1. Calculate the theoretical milliliters of nitrite (Y_c) using Eq. (7.6). This was done in Step 5.1.

Step 2. Subtract the calculated Y values from the observed Y values.

Step 3. Square ($Y_o - Y_c$). This gives the results in Table 7.3.

The quantities $Y_o - Y_c$ represent the deviations of the experimental readings from the line that best fits the data, and hence represent experimental error. The sum of the column should be zero. The sum of the squares of $Y_o - Y_c$ gives the variance with $n - 2$ *df* because the data have been used to estimate both the slope and the intercept.

Table 7.3

X	Y_o	Y_c	$Y_o - Y_c$	$(Y_o - Y_c)^2$
200.0	7.50	7.505	−0.005	0.000025
400.0	14.90	14.885	0.015	0.000225
600.0	22.25	22.265	−0.015	0.000225
800.0	29.65	29.645	0.005	0.000025
			Sum	0.000500

Step 4. Calculate the standard deviation:

$$(n - 2)V = \Sigma(Y_o - Y_c)^2. \tag{7.8}$$
$$2V = 0.0005,$$
$$V_{y/x} = 0.00025,$$
$$s_{y/x} = 0.0158.$$

Step 5. Calculate the standard deviation of the slope $s(b)$:

$$s(b) = s_{y/x}/[\Sigma(X - \bar{X})^2]^{1/2}. \tag{7.9}$$
$$s(b) = 0.0158/(200,000)^{1/2}$$
$$= 0.0158/447.2$$
$$= 0.000035.$$

Step 6. Calculate the standard deviation of the intercept, $s(a)$:

$$s(a) = s_{y/x}\left[\frac{\Sigma X^2}{n\Sigma(X - \bar{X})^2}\right]^{1/2}. \tag{7.10}$$
$$s(a) = 0.0158\left[\frac{1,200,000}{4(200,000)}\right]^{1/2}$$
$$= 0.0158(1.5)^{1/2}$$
$$= 0.0158(1.225)$$
$$= 0.013.$$

Example 7.3 *Tests for Significant Differences*

To determine if the factor is significantly different from 1.00 and if the blank is significantly different from zero, we use the t test.

If the factor were 1.00, the slope would be b'.

$$b' = (1.00 - 0.15)/27.28$$
$$= 0.03116.$$
$$t = (b - b')/s(b)$$
$$= (0.03690 - 0.03116)/0.000035$$
$$= 164.$$

Consulting Table XI for 2 df, this value is much larger than the 1 % critical value for t. We can conclude that the data do not warrant using a factor of 1.00.

The intercept (a) is a measure of the blank titration. Applying the t test:

$$t = (a - 0)/s(a)$$
$$= 0.125/0.013$$
$$= 9.61.$$

This value exceeds the 5 % critical value for t at 2 df and therefore we can conclude that the blank is not caused by experimental error.

7.3 SHORTCUT METHODS

A chemist should not shun the use of regression analysis because of the tedium of the calculations. These computations can be shortened by the use of certain approximate methods and by the proper experimental design. He can control the design by choosing certain points for the X variable.

Example 7.1 was an illustration of this; the weights of Novocain were evenly spaced 200 mg apart and $X_2 = 2X_1$, $X_3 = 3X_1$, and $X_4 = 4X_1$. This makes it possible to use the formulas in Table 7.4.

Table 7.4   Simplified Least Squares Equations for Use When
$X_2 = 2X_1$, *etc.*

n	b	a
2	$\dfrac{Y_2 - Y_1}{X_1}$	$2Y_1 - Y_2$
3	$\dfrac{Y_3 - Y_1}{2X_1}$	$\dfrac{4Y_1 + Y_2 - 2Y_3}{3}$
4	$\dfrac{3Y_4 + Y_3 - Y_2 - 3Y_1}{10X_1}$	$\dfrac{2Y_1 + Y_2 - Y_4}{2}$
5	$\dfrac{2Y_5 + Y_4 - Y_2 - 2Y_1}{10X_1}$	$\dfrac{8Y_1 + 5Y_2 + 2Y_3 - Y_4 - 4Y_5}{10}$

These formulas have been worked out for cases where $X_n = nX_1$.
If $X_1 = 0$, as in the case of some blanks in colorimetric analysis,
Table 7.5 is used. In this table, $X_n = (n - 1)X_2$.

Table 7.5   For Use When $X_1 = 0$, $X_3 = 2X_2$, *etc.*

n	b	a
2	$\dfrac{Y_2 - Y_1}{X_2}$	Y_1
3	$\dfrac{Y_3 - Y_1}{2X_2}$	$\dfrac{5Y_1 + 2Y_2 - Y_3}{6}$
4	$\dfrac{3Y_4 + Y_3 - Y_2 - 3Y_1}{10X_2}$	$\dfrac{7Y_1 + 4Y_2 + Y_3 - 2Y_4}{10}$
5	$\dfrac{2Y_5 + Y_4 - Y_2 - 2Y_1}{10X_2}$	$\dfrac{3Y_1 + 2Y_2 + Y_3 - Y_5}{5}$

Another shortcut is the use of range in place of standard deviation for tests of significance. Both these shortcut methods are illustrated in the following example.

Example 7.4

Using the data in Example 7.1

$$X_1 = 200, \quad Y_1 = 7.50;$$
$$X_2 = 400, \quad Y_2 = 14.90;$$
$$X_3 = 600, \quad Y_3 = 22.25;$$
$$X_4 = 800, \quad Y_4 = 29.65.$$

Substituting these values in the equations for $n = 4$, in Table 7.4.

$$b = \frac{3Y_4 + Y_3 - Y_2 - 3Y_1}{10X_1}.$$

$$b = \frac{3(29.65) + 22.25 - 14.90 - 3(7.50)}{10(200)}$$

$$= 0.03690.$$

$$a = \frac{2Y_1 + Y_2 - Y_4}{2}.$$

$$a = \frac{2(7.50) + 14.90 - 29.65}{2}$$

$$= 0.125.$$

These values are substituted in Eq. (7.2)

$$Y = 0.125 + 0.03690X.$$

When we substitute in Eq. (7.2) we obtain Y_c: the calculated value of Y. Comparing Y_c with the observed values of the amount of sodium nitrite Y_o we obtain results as shown in the following tabulation.

X	Y_o	Y_c	d
200	7.50	7.505	0.005
400	14.95	14.885	0.015
600	22.20	22.265	0.015
800	29.65	29.645	0.005
			$\Sigma d = 0.040$

The distance in either direction from Y_c, the center line, Σd, is only half the range.

$$R_{y/x} = 2\Sigma d. \qquad (7.11)$$
$$R_{y/x} = 2(0.04)$$
$$= 0.08.$$

From Eq. (7.9) we can derive:

$$R_b = \frac{R_{y/x}}{[\Sigma(X - \bar{X})^2]^{1/2}}. \qquad (7.12)$$
$$R_b = \frac{0.08}{447.2}$$
$$= 0.00018.$$

The following equation can be derived from Eq. (7.10):

$$R_a = R_{y/x}\left[\frac{\Sigma X^2}{n\Sigma(X - \bar{X})^2}\right]^{1/2}. \qquad (7.13)$$
$$R_a = 0.08(1.225)$$
$$= 0.098.$$

Tests of significance are carried out using the L test.

To test if the factor is significantly different from 1.00, we again test the difference between the calculated slope and the theoretical slope.

$$L = \frac{b - b'}{R_b}. \tag{7.14}$$

$$L = \frac{0.03690 - 0.03116}{0.00018}$$

$$= 31.9.$$

Referring to Table III, we see the difference is highly significant. To test if the blank titration is significantly different from zero:

$$L = \frac{a - o}{R_a}. \tag{7.15}$$

$$L = 0.125/0.098$$

$$= 1.28.$$

This difference is also highly significant.

7.4 SHORTCUT METHOD WHEN X_n DOES NOT EQUAL nX_1

There are occasions when it is impossible for the analyst to choose the points of the independent variable, in which case X_n may not be a multiple of X_1, and the equations in Tables 7.4 and 7.5 will not apply. It may still be possible to use a shortcut procedure. When there are two or three observations the following formulas can be applied:

$$b = \frac{Y_n - Y_1}{X_n - X_1}. \tag{7.16}$$

$$a = \bar{Y} - b(\bar{X}). \tag{7.17}$$

The use of these equations is illustrated in Example 7.9.

7.5 COLORIMETRIC ANALYSIS

Colorimetric assays involve the preparation of a series of standard solutions and the comparison of the intensity of the color with that produced simultaneously in a solution of unknown concentration. The concentration of the unknown is interpolated from the relation of its absorbence to the absorbencies of the standard solutions in a colorimeter or spectrophotometer. The relationship of absorbence to concentration is linear (Beer's law), and as such follows the laws of the straight line as discussed earlier.

In colorimetry the concentration of the substance being measured is the X variable, the absorbence is the Y variable. The slope is a measure of the change in absorbence for a unit change in concentration. The intercept (a) is a measure of the blank.

There are two techniques in general use in colorimetry. One involves the interpolation of the unknown from a standard curve. The other technique makes use of the ratio between the unknown and a standard, at approximately the same concentration. The hypothesis is that the slope of the unknown curve equals the slope of the standard curve:

$$b_u = \frac{Y_u - a_u}{X_u} = b_s = \frac{Y_s - a_s}{X_s},$$

$$X_u = \frac{X_s(Y_u - a_u)}{Y_s - a_s}. \tag{7.18}$$

Often, because of the design of the instrument or the nature of the reaction, the blank is zero and Eq. (7.18) becomes:

$$X_u = \frac{X_s Y_u}{Y_s}.$$

The following examples have been worked out, using data from an example in "Colorimetric Methods of Analysis" by Snell and Snell (*1*). The data are taken unaltered from the original.

Original data	Standard curve
X Silica (mg)	Y Absorbence
0	0.032
0.02	0.135
0.04	0.187
0.06	0.268
0.08	0.359
0.10	0.435
0.12	0.511
Unknown	$Y = 0.242$

The authors find a silica content of 0.053 mg in the sample, using a graphical method, and 0.052 mg using an arithmetical method. They give no estimate of the precision of either result.

When we use regression analysis as outlined in Section 7.2 we obtain the following results:

$$a = \text{blank} = 0.039 \pm 0.008 \quad (95\% \text{ CL}),$$
$$X = 0.052 \pm 0.002 \qquad\qquad (95\% \text{ CL}).$$

The analysis requires making seven standard and one unknown color reactions. The next few examples will be used to illustrate shorter methods.

Example 7.5

If the analyst runs a blank, a standard at 0.06 mg, and the unknown, he will have run three determinations. If he assumes the blank is the same for both standard and unknown, he can use:

$$X_u = \frac{X_s(Y_u - a_u)}{Y_s - a_s} . \qquad (7.19)$$

$$X_u = \frac{0.06(0.242 - 0.032)}{0.268 - 0.032}$$

$$= 0.053 \quad \text{mg}.$$

He does not have enough data to estimate the experimental error.

Example 7.6

Suppose the analyst runs a blank, a standard at 0.02 and at 0.04 mg, and the unknown. His results will be:

$$X_1 = 0.00, \qquad Y_1 = 0.032;$$
$$X_2 = 0.02, \qquad Y_2 = 0.135;$$
$$X_3 = 0.04, \qquad Y_3 = 0.187.$$
$$\Sigma X^2 = 0.002;$$
$$\Sigma(X - \bar{X})^2 = 0.0008.$$

Substituting these data in the equations for $n = 3$ in Table 7.5:

$$b = \frac{Y_3 - Y_1}{2X_2}$$

$$= \frac{0.187 - 0.032}{2(0.02)}$$

$$= 3.88.$$

$$a = \frac{5Y_1 + 2Y_2 - Y_3}{6}$$

$$= \frac{5(0.032) + 2(0.135) - 0.187}{6}$$

$$= 0.040.$$

Substituting in Eq. (7.2) and solving for X:

$$X = \frac{0.242 - 0.040}{3.88} = 0.052 \quad \text{mg.}$$

This experiment has enough data to enable the analyst to estimate the confidence limits of (*a*) the blank, and (*b*) the analytical result.

The range of the regression is computed by calculating the differences between Y_o and Y_c as in Section 7.3

X	Y_o	Y_c	d
0.00	0.032	0.040	0.008
0.02	0.135	0.118	0.017
0.04	0.187	0.195	0.008
			$\Sigma d = 0.033$

$$R_{y/x} = 2(0.033) = 0.066.$$

Substituting in Eq. (7.13):

$$R_a = 0.066 \left[\frac{0.002}{3(0.0008)} \right]^{1/2}$$

$$= 0.060.$$

Testing the null hypothesis $a = 0$:

$$L = \frac{0.040 - 0}{0.060} = 0.67.$$

Consulting Table III, at $n = 3$ we find the critical value for $L_{0.05}$ to be 1.30.

Example 7.7

If the analyst had run the unknown, and standard determinations at 0.02, 0.04, and 0.06 mg, he would have

$$X_1 = 0.02, \qquad Y_1 = 0.135;$$

$$X_2 = 0.04, \qquad Y_2 = 0.187;$$

$$X_3 = 0.06, \qquad Y_3 = 0.268.$$

$$\Sigma X^2 = 0.0056;$$

$$\Sigma(X - \bar{X})^2 = 0.0008.$$

In this example, X_1 does not equal zero, so the formulas in Table 7.4 should be used:

$$b = \frac{Y_3 - Y_1}{2(X_1)}$$

$$= \frac{0.268 - 0.135}{2(0.02)} = 3.325.$$

$$a = \frac{4Y_1 + Y_2 - 2Y_3}{3}$$

$$= \frac{4(0.135) + 0.187 - 2(0.268)}{3} = 0.064.$$

Substituting Eq. (7.2) and solving for X:

$$X = \frac{0.242 - 0.064}{3.325} = 0.054.$$

Example 7.8

If the analyst runs a blank, the unknown, and standards at 0.06 and 0.12, he will have:

$$X_1 = 0.00, \qquad Y_1 = 0.032;$$

$$X_2 = 0.06, \qquad Y_2 = 0.268;$$

$$X_3 = 0.12, \qquad Y_3 = 0.511.$$

$$\Sigma X^2 = 0.018;$$

$$\Sigma(X - \bar{X})^2 = 0.0072.$$

Substituting in the appropriate equations in Table 7.2:

$$b = \frac{0.511 - 0.032}{2(0.06)} = 3.98.$$

$$a = \frac{5(0.032) + 2(0.268) - 0.511}{6} = 0.031.$$

These values are substituted in Eq. (7.2) and the equation is solved for X:

$$Y = 0.242 - 0.031 + 3.98X.$$

$$X = \frac{0.242 - 0.031}{3.98} \, ;$$

$$X = 0.053 \quad \text{mg.}$$

This is the result obtained in the original article by graphical methods.

The analyst needs to know how much confidence he can place in his blank.

The range of the regression is computed by calculating the differences between Y_o and Y_c, as in Section 7.3

X	Y_o	Y_c	d
0.00	0.032	0.031	0.001
0.06	0.268	0.270	0.002
0.12	0.511	0.509	0.002
			$\Sigma d = 0.005$

$$R_{y/x} = 2(0.005)$$
$$= 0.01 \, ;$$

$$R_a = R_{y/x} \left[\frac{\Sigma X^2}{n\Sigma(X - \bar{X})^2} \right]^{1/2}$$

$$= 0.009.$$

Testing the null hypothesis, $a = 0$:

$$L = \frac{0.031 - 0}{0.009}$$

$$= 3.40.$$

Consulting Table III, we find that there is a very high probability that a does not equal zero.

7.6 CONFIDENCE LIMITS FOR X

Although the weights of the standards are dependent variables, the calculated response depends upon the colorimetric reading which is a dependent variable and therefore does have an analytical error associated with it. The error is a function of the response and the slope and can be approximated by

$$s_x = (s_{y/x})/b,$$

or

$$R_x = (R_{y/x})/b. \qquad (7.20)$$

We now can estimate confidence limits for the results obtained in the preceding examples

Example	$R_{y/x}$	b	R_x
7.5	None	—	—
7.6	0.066	3.88	0.017
7.7	0.038	3.325	0.011
7.8	0.01	3.98	0.0025

Table 7.6   Summary

Original data	N	a	CL	X (mg)	95% CL
Graphic solution	8	0.032	None	0.053	None
Regression solution	8	0.039	±0.008	0.052	±0.002
Example 7.5	3	0.032	None	0.054	None
Example 7.6	4	0.040	±0.086	0.052	±0.012
Example 7.7	4	0.064	±0.049	0.054	±0.008
Example 7.8	4	0.031	±0.013	0.053	±0.002

Examples 7.6 and 7.7 have a large error, whereas Example 7.8 has an error comparable to that obtained when twice as many points are run. This illustrates an important feature of regression analysis; a wide spacing of the points of the standard improves the precision.

7.7 NONLINEAR FUNCTIONS

It is sometimes possible to treat nonlinear data in the manner described above by a simple transformation. For example, in pharmacology the response of an animal to a drug is often a function of the log of the dose: $Y = a + \log X$; hence, plotting Y against $\log X$ gives a straight line. Similarly, the rate of a chemical reaction can be plotted as a straight line against time by use of the following transformations: monomolecular reaction, $Y = \log C$; bimolecular reaction, $Y = 1/C$; trimolecular reaction, $Y = 1/C^2$, etc.

Example 7.9 *The Study of a Monomolecular Decomposition*

The hydrolysis of a solution of a certain organic compound yields a product which is relatively insoluble. When 1% of the hydrolysis product is present in the solution, crystals form which render the solution unsalable. An experiment was designed to predict the shelf life of the product.

The product was filled into commercial containers, sterilized, and kept at constant temperature for seven months. Samples were analyzed for the percentage of hydrolyzed product at the end of 3, 5, and 7 months, with the results given below. The hydrolysis was a monomolecular reaction; consequently the reaction was studied in the following form: log of the concentration is equal to a function of time; i.e., $\log Y = a + bX$.

Age of solution (months)	Hydrolysis (%)
3	0.220
5	0.255
7	0.304

The analysis of the data proceeds as follows:

Step 1.   Convert the percentage of hydrolysis into log of ten times the percentage of hydrolysis. This is Y_o in Table 7.7.

Table 7.7

X	$Y_o = \log 10C$	Y_c	d
3	0.3424	0.3404	0.0020
5	0.4065	0.4106	0.0041
7	0.4829	0.4808	0.0021
$\bar{X} = 5$	$\bar{Y} = 0.4106$		$d = 0.0082$
$\Sigma \bar{X}^2 = 83$			
$\Sigma(X - \bar{X})^2 = 4$			

Step 2. The values of X are not multiples of each other. In this case, we use Eqs. (7.16) and (7.17) to calculate b and a:

$$b = \frac{Y_n - Y_1}{X_n - X_1}$$

$$= \frac{0.4829 - 0.3424}{7 - 3} = 0.0351.$$

$$a = \bar{Y} - b\bar{X}$$

$$= 0.4106 - 0.0351(5) = 0.2351.$$

Step 3. Substitute in Eq. (7.2)

$$Y_c = \log 10C = 0.2351 + 0.0351 X.$$

We calculate Y_c from the above equation. The Y_c values are tabulated in Column 3, Table 7.7.

$$\Sigma d = \Sigma(Y_o - Y_c) = 0.0082.$$

$$R_{y/x} = 2\Sigma d = 0.0164;$$

$$R_a = R_{y/x} \left[\frac{\Sigma X^2}{n\Sigma(X - \bar{X})^2} \right]^{1/2}$$

$$= 0.0164 \left[\frac{83}{3 \times 4} \right]^{1/2}$$

$$= 0.0431.$$

To test if a is real:

$$L = a/R_a$$

$$= 0.2351/0.0431$$

$$= 5.45.$$

This is highly significant. Hence, a is real.

a is the log of ten times the initial concentration of hydrolyzed material. Since it does not equal zero, we must conclude that there is an initial hydrolysis of 0.172% caused by the sterilization of the solution.

Crystals will form when the hydrolysis reaches 1%. The time at which this can be expected to occur can be calculated as follows:

$$\text{log of ten times } 1\% = 1.0000;$$

$$X = \frac{1.0000 - 0.2351}{0.0351} = 21.8 \quad \text{months};$$

$$R_x = \frac{R_{y/x}}{b}$$

$$= \frac{0.0164}{0.0351}$$

$$= 0.47.$$

The 95% confidence limits are:

$$21.8 \pm 2.25(0.47) = 20.7\text{--}22.9$$

We can expect this solution to have a shelf life of slightly less than 2 years.

REFERENCE

1. Snell, F. D., and Snell, C. (1951). "Colorimetric Methods of Analysis," 3rd ed., Vol. 1, p. 151. Van Nostrand-Reinhold, Princeton, New Jersey.

8

Sampling

8.1 THE SAMPLE AND THE POPULATION

In chemical work, we take samples from a batch of material, test them, and make deductions about the whole batch from the results of these tests.

It is axiomatic that the best analysis is only as good as the sample which is tested. The sample is not the population but is presumed to represent the population and the validity of any conclusions drawn from it depends upon how truly representative it is.

The relationship between sample and population can be illustrated in the following way:

Suppose a person who has never seen a deck of playing cards is asked to describe it from a sample. The cards are shuffled and he draws a single card—let us say it is the nine of clubs. The only thing he knows from this sample is that one card has nine black cloverlike figures on it. This information may lead him to assume:

(1) All cards have nine black cloverlike figures.
(2) The cards are numbered differently and decorated with cloverlike figures.

Both assumptions are false because the sample is too small to give an indication of the heterogeneous nature of the population.

If he draws five cards, he may draw:

Spades	K, 6
Hearts	K, 7
Diamonds	—
Clubs	3

This sample is more informative. He now knows:

(1) There are at least two different colors—red and black.
(2) There are at least three different suits in the pack.
(3) The cards have pictures as well as numbers on them.
(4) There is a duplication of pictures in different suits.

The facts he now has may lead him to postulate:

(1) The existence of a fourth suit in order to give symmetry to the deck.
(2) The missing fourth suit is probably red.
(3) Since there is a duplication of picture cards, there probably is a duplication of number cards.

If he draws thirteen cards, he may draw a sample of this nature:

Spades	8
Hearts	K, 10, 8, 7, 2
Diamonds	A, 9, 7, 6, 2
Clubs	5, 4

He now has evidence that his assumptions from a five-card sample are true. He may now be reasonably sure:

(1) There are only four suits in a deck of cards.
(2) There are two, and only two, colors.
(3) The cards are numbered consecutively from two to ten, and the "one" card is called "A."
(4) Cards with pictures are relatively scarce and therefore probably have a premium value.
(5) There has been no duplication of numbers in the same suit, and hence probably none exists.

An examination of the nineteen cards he has drawn in the three samples would give him added assurance of these points, and he could now give the following description:

A pack of cards is divided into four suits, distinguished by symbols. Two suits have red symbols, and two have black symbols. Each suit is consecutively numbered, A (1) through 10, and K. There are, therefore, at least eleven cards in each suit, or a total of at least 44 cards in the deck.

Much of this information was obtained, or could have been surmised, when a sample of five cards was used. The larger sample and pooled information of the three samples proved the assumptions made after examination of five cards. By a quirk of chance, none of the samples disclosed the presence of a Jack or Queen, and so the man could not have predicted the correct number of cards in the deck. This is one of the risks involved in any sampling scheme.

There is one more point to be made from the deck of cards example. If the man had taken a sample of thirteen cards from a fresh pack without shuffling them, he would have come to the conclusion that all cards were of the same color and suit, that they were numbered consecutively from 1 to 10, followed by a series of pictures. In this case, the sample was not random, and although it would have given a complete picture of one suit, the man could not have guessed the presence of the other three suits without a sample size of 40 cards.

The picture changes sharply if the man knows something about a pack of cards. It is this knowledge of the population which enables a person to make a decision when playing card games. A hand dealt to a player is really a sample, and the player's ability to judge the strength of his sample in respect to his opponents' samples is what is often called "card sense."

This is a rough analogy between research and control. The research chemist is often in the position of the man who knows nothing about the population he is sampling. The control chemist is at the mercy of the laws of probability as they govern samples, but he usually has some prior knowledge on which to base his judgments.

8.2 THE THEORY OF SAMPLING

From the preceding general description of a sampling experiment it is evident that certain risks are involved. Specifically, these risks are of two kinds: (*i*) the risk of accepting a bad lot of material, and (*ii*) the risk of rejecting a good lot of material. These risks exist independently of human mistakes in analyzing the sample; a bad lot may give a good sample and a good lot may give a bad sample.

For obvious reasons, we wish to keep these risks at a minimum. The only way of reducing them to zero is to test the whole lot. In cases where testing is expensive this is economically unsound, and in cases where the testing is destructive we would collect much interesting data, but we would have nothing to sell.

Good and bad are relative terms as used above. We would like to be able to define a good lot as one with no defective pieces, and a bad lot as one with one or more defectives. However, this is not practical short of 100% inspection; if we were to test 99 pieces out of a lot of 100 and found them all perfect, we could not be certain that the one-hundredth piece was not defective. The only practical approach is to realize this and to tolerate the possibility of a fraction of defective material. Then we can define a good lot as one that contains no more than this fraction defective (designated by p_1). It would be ideal if we could say that a bad lot is one that contains more than p_1 fraction defective, but unfortunately this is impossible. A lot with $p_1 = 0.11$ is only slightly better than one with $p_1 = 0.13$, and it would take a very large sample to tell the difference. For practical purposes, we must choose another value of fraction defective (say p_2), larger than p_1, to define a bad lot. The closer p_1 and p_2 are to each other, the better discrimination we have between good and bad lots. The only way to obtain this sharp discrimination is to use large samples ($p_1 = p_2$ at 100% testing).

If we plot the probability of accepting the lot against the fraction of defective material in the lot, we obtain a curve which is characteristic for each sampling plan. This curve is known as the operating characteristic (OC) curve. If we choose $p_1 = 1\%$, $p_2 = 5\%$ with 5% risks, the probability of accepting a lot with 1% defective material would be 0.95; the probability of accepting a lot with 5% defective would be 0.05, and the OC curve would, of course, pass through both points.

Statistically speaking, there are two types of sampling: (*i*) by attributes (counting data), and (*ii*) by variables (measurement data).

It is the former that most often interests the bacteriologist (bacteria either live or die), and the latter that most often interests the chemist. Sampling by variables is more efficient than sampling by attributes; that is, it takes smaller samples to discriminate between good and bad lots.

8.3 SAMPLE SIZE

The chemist, especially the analytical chemist, frequently asks the question, "What size sample shall I use?" As it stands, this is not a question so much as it is an exclamation—often of despair. What chemists really mean is, "What is the most economical sample size we can take and still be reasonably sure that the deductions we draw from the sample are valid for the population?" Before answering this question we must first determine:

(1) What type sampling are we going to use, attribute or variable?
(2) What fraction defective is acceptable (p_1)?
(3) What fraction defective is unacceptable (p_2)?
(4) What risk can we take of rejecting good material?
(5) What risk can we take of accepting bad material?

The answer to the first question is fixed by the problem; certain kinds of testing cannot give data which can be treated as a variable.

The answers to the last four questions are a matter of economics or of moral obligation to a customer. If the defect is a minor one entailing, for example, a matter of appearance which does not influence the functioning of the product, greater risks and wider limits can be tolerated. If, on the other hand, the appearance does affect the salability or the functioning of the product, we naturally have to work to closer tolerances and lower risks of making a wrong decision. The moral obligation obtains in the case of contractual relationships or, as in the pharmaceutical industry, where the purity of the product may be a matter of life or death.

Prior knowledge of the population is a help in deciding sample size. If we know we are sampling a homogeneous population, such as a well mixed solution, we know that one sample taken from anywhere will give as much information as ten samples taken from ten different parts of the solution. On the other hand, if we suspect or know that the material being sampled is not homogeneous, we must sample from various parts of the lot. Prior knowledge regarding the sampling error permits the analyst to work with smaller samples.

8.4 ATTRIBUTE SAMPLING

Attribute sampling means inspecting a sample for some attribute and accepting or rejecting the lot on the basis of the number of good or bad pieces in the sample. Statisticians have studied attribute sampling and many prepared sampling plans are available.

Pioneers in sampling, H. F. Dodge and H. G. Romig (*1*), prepared sampling plans for use in the Bell Telephone System in 1929. These plans were based on the concept of average outgoing quality limit (AOQL). In an AOQL plan a sample is drawn and inspected. If the sample passes, the lot is accepted. If the sample fails, the lot is subjected to 100% sorting. In the long run lots of material subjected to this type of inspection will have a quality level better than the specified AOQL.

Dodge and Romig also developed sampling plans based on the lot tolerance percent defective (LTPD). The LTPD is lot quality for which the probability of acceptance is 10%; that is to say, the consumer's risk (p_2) is 0.10.

Probably the most commonly used sampling scheme is MIL-STD-105D (Military Standard 105D) (*2*). It was developed by a working group of American, British, and Canadian statisticians and hence is a common standard for all three countries. It is described in the publication, "Military Standard 105D Sampling Procedures and Tables for Inspection by Attributes."

MIL-STD-105D is an inspection system consisting of a collection of sampling plans, each of which is listed according to its Acceptable Quality Level (AQL).

AQL is defined as, "the maximum percent defective that, for the purpose of sampling inspection, can be considered satisfactory as a process average"(2). The satisfactory maximum percentage defective is not always the same for plans with the same AQL, but may vary from about 86% to about 97%.

The plans in MIL-STD-105D are designed to give the vendor protection against the rejection of lots from a process that is producing at the AQL or better. However, it may give the consumer unsatisfactory protection against accepting lots that are moderately worse than the AQL. The standard compensates for this possibility by providing for more severe inspection whenever the quality history of the product is bad or doubtful.

MIL-STD-105D also recognizes differences in the severity of defects and defines and classifies them as "Critical," "Major," and "Minor," permitting different AQL levels for each classification.

Figure 8.1 shows the operating characteristic curves for two different sampling plans from 105D. Both plans are characterized as AQL = 1.0% plans. Plan E calls for a sample size of 13 and accepting the lot if no defects are found in the sample. If, on the average, lots of 1% defective were submitted, they would be accepted 85% of the time. Plan J, which calls for a sample size of 80 and permits two defects in the sample, would accept 1% defective lots about 96% of the time. Plans E and J are equivalent at 2.2% defective. Plan E will accept 10% defective 26% of the time; Plan J practically never. Note that Plan E will accept lots with 16% defective about 10% of the time.

Both plans are given in 105D for inspection of the same lot size (500–1200). E is Level S3—used for expensive or destructive testing; J is the Level II—the normal level of inspection.

8.5 SAMPLING BY VARIABLES

Ideally, when decisions are to be made upon measurement data, sampling schemes based upon variables should be used since variables sampling plans require smaller sample sizes than do attribute plans affording equal protection. Unfortunately, practically all existing variables sampling plans are predicated upon the assumption that the sample comes from a Normally Distributed Population. The United States Department of Defense has published MIL-STD-414 (*3*) based on the work of Lieberman and Resnikoff (*4*). Both assume normality.

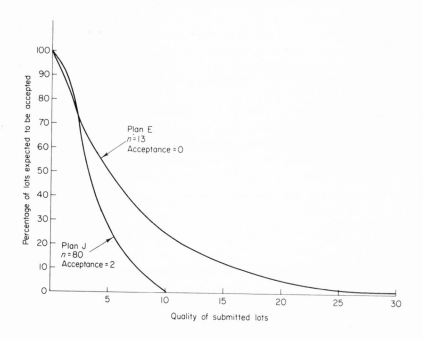

Figure 8.1. Operating characteristic curves for two different sampling plans from 105 D. Both plans are characterized as AQL = 1.0% plans.

When there is serious doubt as to the normality of the parent distribution it is safer to rely upon attribute sampling. This produces a plan in which a piece outside of specification tolerances is the defect.

8.6 USE OF COMPONENTS OF VARIANCE

In general, the variance (and hence the precision) of an analytical procedure depends upon the variance of the sample and the variance of the analysis:

$$V_t = \frac{V_s}{k} + \frac{V_a}{kn}, \qquad (8.1)$$

where k is the number of samples and n is the number of analyses per sample. The product kn is the total number of analyses. The analyst wishes to minimize V_t to obtain the best possible precision, and also to minimize kn to do the least amount of testing. Proper selection of the sample will accomplish this.

In Chapters 3 and 5, methods were given for isolating and estimating the components of variance of an experimental design. The EMS components are a measure of the variance contributed by the factors being studied.

If we have an EMS for samples of 0.0593 and an EMS for error of 0.04, and we take one sample from anywhere in the lot and analyze it, the total variance of the analysis is given by substituting in Eq. (8.1).

$$V_t = \frac{0.0593}{1} + \frac{0.04}{1 \times 1}$$

$$= 0.0993.$$

$$s_t = 0.315.$$

If we run duplicate analyses, the total variance is:

$$V_t = \frac{0.0593}{1} + \frac{0.04}{2}$$

$$= 0.0793.$$

$$s_t = 0.28.$$

This represents a slight improvement in precision, but it is hardly worth the extra work. Proper sampling, however, will effect an appreciable improvement, as Table 8.1 demonstrates.

Table 8.1

k	n	kn	V_t	s_t
1	1	1	0.0993	0.32
1	2	2	0.0793	0.28
1	4	4	0.0693	0.26
2	1	2	0.0396	0.20
4	1	4	0.0248	0.16

Table 8.1 shows that taking a single sample and running duplicate or even quadruplicate analyses does little to improve the precision, whereas choosing two or four samples and analyzing each increases the precision appreciably.

Example 8.1

In Example 5.7, we made a study of the effect of ovens and trays in ovens on a drying process, and found that the EMS of the factors was as shown in the next tabulation.

	EMS	s
V_e	0.1149	0.34
V_t	0.0309	0.17
V_o	0.0098	0.10

To maximize the precision with a minimum of testing we must reduce V_o to a minimum. Let us examine some of the possible ways of doing this:

(1) Take a single sample and analyze it:

Total $V = 0.1149 + 0.0309 + 0.0098 = 0.1556;$

$s = 0.395.$

(2) Run duplicate analyses on a single sample:

Total $V = 0.1149 + 0.0309 + \dfrac{0.0098}{2} = 0.1507;$

$s = 0.388.$

(3) Take a sample from each of two trays in an oven and analyze each:

Total $V = 0.1149 + \dfrac{0.0309 + 0.0098}{2} = 0.1352;$

$s = 0.368.$

(4) Take a sample from each oven and analyze each:

Total $V = \dfrac{0.1149 + 0.0309 + 0.0098}{2} = 0.0778;$

$s = 0.279.$

The only practical reduction in variance is achieved by Plan 4, which increases the precision by about 30% over the other plans.

8.7 VARIABLES PLAN BASED UPON NORMAL DISTRIBUTION

If we know the standard deviation of the method we can determine sample size by use of the normal distribution. The distance of an average $\bar{X}$ from the true mean μ is a constant times the standard deviation:

$$\bar{X} - \mu = k s_{\bar{X}},$$

where $s_{\bar{X}} = s/\sqrt{n}$ and the constant k is the sum of the risk probabilities

$$k = k_1 + k_2$$

so that

$$\bar{X} - \mu = \frac{(k_1 + k_2)s}{\sqrt{n}}.$$

Solving for n,

$$n = \frac{(k_1 + k_2)^2 s^2}{(\bar{X} - \mu)^2}. \qquad (8.2)$$

If we take equal risks, Eq. (8.2) can be simplified to give the equations shown in the following tabulation:

Risk	Single limit test		Double limit test	
0.05	$\dfrac{10.8(s)^2}{(\bar{X} - \mu)^2}$	(8.3)	$\dfrac{13.0(s)^2}{(\bar{X} - \mu)^2}$	(8.5)
0.01	$\dfrac{21.6(s)^2}{(\bar{X} - \mu)^2}$	(8.4)	$\dfrac{24.0(s)^2}{(\bar{X} - \mu)^2}$	(8.6)

The use of these equations is illustrated by the following examples:

Example 8.2 *Single Limit Test*

The specifications call for a product to be not less than 95% pure. We know the standard deviation of the method is 3.1%. The problem is to design a sampling plan which will give 95% assurance that the product meets the specifications.

$$\bar{X} - \mu = 95 - 100 = -5;$$

$$s = 3.1.$$

Substituting in Eq. (8.3) we have,

$$n = \frac{10.8(3.1)^2}{(-5)^2}$$

$$= 4.2.$$

Use a sample of 5 and accept if the average is greater than X_0 as defined by

$$X_0 = \frac{\bar{X} + \mu}{2}; \qquad (8.7)$$

$$X_0 = \frac{100 + 95}{2} = 97.5\%.$$

If we desire to be 99% sure of our results, we substitute in Eq. (8.4):

$$n = \frac{21.6(3.1)^2}{(-5)^2} = 8.3.$$

This plan calls for nine analyses.

Example 8.3 *Double Limit Test*

The specifications call for a product to contain not less than 95% nor more than 105% of the label claim. We know the standard deviation of the method is 3.1%. Design a sampling plan that will give (a) 95% assurance, (b) 99% assurance that the product will meet these requirements.

$$(\bar{X} - \mu) = 105 - 100 = 5;$$

$$s = 3.1.$$

(a) Substitute in Eq. (8.5):

$$n = \frac{13.0(3.1)^2}{(5)^2} = 5.$$

Ninety-five percent limits are given by

$$d = \frac{1.96(s)}{\sqrt{n}} ; \qquad (8.8)$$

$$d = \frac{1.96(3.1)}{\sqrt{5}} = 2.72.$$

The limiting values are

$$X_0 = \mu \pm d. \tag{8.9}$$

Therefore, the plan calls for running five analyses and accepting if the average lies between 97.28 and 102.72%.

(b) Substitute in Eq. (8.6)

$$n = \frac{24(3.1)^2}{(5)^2} = 9.2.$$

Ninety-nine percent limits are

$$d = \frac{2.58(s)}{n}; \tag{8.10}$$

$$d = \frac{2.58(3.1)}{10};$$

$$d = 2.53.$$

This value for d is substituted in Eq. (8.9) to obtain the limiting values.

If we want 99% assurance that the specifications will be met, we should run ten determinations and accept the lot if the average falls between 97.47 and 102.53%.

REFERENCES

1. Dodge, H. F., and Romig, H. G. (1929). A method of sampling inspection. *Bell Syst. Tech. J.* **8**, 613. See also Dodge, H. F. (1969). *J. Quality Technol.* **1**, 77, 155, 255; **2**, 1.
2. MIL-STD-105D (1963). Sampling procedures and tables for inspection of attributes. Department of Defense, Washington, D.C.
3. Lieberman, G. J., and Resnikoff, G. J. (1955). *J. Amer. Statist. Assoc.* **50**, 457.
4. Military Standards 414 (1957). Sampling procedures and tables for inspection by variables for percent defective. U.S. Govt. Printing Office, Washington, D.C.

9

Control of Routine Analysis

9.1 PROBLEMS OF THE ROUTINE ANALYST

The purpose of this chapter is to show how statistical methods can be applied to the problems of the routine analyst. These problems are sometimes solved by rule of thumb, sometimes by intuition (which is often highly developed in an experienced analyst), and sometimes left unsolved. At best these are subjective methods; the statistical approach introduces objectivity into the solution.

The primary job of the routine analyst is to determine if a product meets specifications. Analytical procedures can be considered to be designed experiments in which conditions have been prescribed to control as many variables as possible.

However, a complete control of such factors as samples, differences between analysts, instrument responses and the like is often impossible; it is these variables which make up analytical "error."

Analytical variability affects the precision and the accuracy of our results, and consequently influences our judgments. It is meaningless to say an analysis is "within the limits of experimental error" if we have no idea of the magnitude of the experimental error.

Many of the illustrations will make use of the data in Table 9.1 which has been taken from "A Study of the Accuracy and Precision of Microanalytical Determinations of Carbon and Hydrogen" by F. W. Powers, published in *Industrial and Chemical Engineering, Analytical Edition* in 1939.

The two columns represent individual carbon analyses by two different analysts.

Table 9.1 Percentage of Carbon in Ephedrine Hydrochloride

	Analyst H	Analyst FWP
	59.09	59.51
	59.17	59.75
	59.27	59.61
	59.13	59.60
	59.10	—
	59.14	—
$\bar{X}$	59.15	59.62
R	0.18	0.24
n	6	4
s	0.065	0.099

9.2 TEST FOR OUTLIERS

Given these figures, the analysts ask themselves a number of questions, the first of which would probably be: "Are these figures homogeneous—that is, do they show any abnormal variations?"

Intuitively, Analyst H could feel comfortable; there are no gaps larger than 0.10%. Analyst FWP might be tempted to discard his result of 59.75% since it is higher than the theoretical value for carbon in ephedrine hydrochloride, (59.55%), and since it is 0.14% higher than the next lower value.

The statistical approach would be to use a test for outliers. Such a test was described in Section 2.8.

Analyst H	Analyst FWP
$\bar{X} = 59.15$	$\bar{X} = 59.62$
$R = 0.18$	$R = 0.24$
$n = 6$	$n = 4$
Extreme value 59.27	Extreme value 59.75
$t_i = \| 59.27 - 59.15 \| /0.18$	$t_i = \| 59.75 - 59.62 \| /0.24$
$= 0.667$	$= 0.542$

Consulting Table X, we find that both values of t_i are less than the critical values, hence there are probably no abnormal values.

9.3 PRECISION OF THE ANALYSES

Both analysts have a range of about 0.2%, which is about 2 parts in 600, so that they may estimate the precision at about 0.33%.

Confidence limits can be calculated by either the range or the standard deviation:

$$CL = \bar{X} \pm (ts/\sqrt{n}), \tag{9.1}$$

where $s = [(\Sigma X - \bar{X})^2/(n - 1)]^{1/2}$

Example 9.1 *Precision Using Standard Deviation*

For Analyst FWP

X	$X - \bar{X}$	$(X - \bar{X})^2$
59.51	-0.09	0.0081
59.75	0.13	0.0169
59.61	-0.01	0.0001
59.60	-0.02	0.0004
$\bar{X}$ 59.62		0.0255

$$s = [0.0255/3]^{1/2} = 0.099$$

For Analyst H, $s = 0.065$.

For Analyst FWP,

$$95\% \ CL = 59.62 \pm [3.18(0.099)/\sqrt{4}]$$
$$= 59.46 \text{ to } 59.78\%.$$

For Analyst H,

$$95\% \ CL = 59.15 \pm [2.57(0.065)/\sqrt{6}]$$
$$= 59.08 \text{ to } 59.22\%.$$

Example 9.2 *Precision Using Range*

One can calculate confidence limits from the range by using the factor A in Table I.

$$CL = \bar{X} \pm A\Sigma R.$$

For Analyst H,

$$CL = 59.15 \pm [0.402(0.18)] = 59.08 \text{ to } 59.22\%.$$

For Analyst FWP,

$$CL = 59.62 \pm [0.735(0.24)] = 59.44 \text{ to } 59.80\%.$$

9.4 DIFFERENCE BETWEEN ANALYSTS

The two analysts differ by 0.47—that is, by about 5 parts in 600. This is less than 1% and hence intuitively they might believe the difference is not significant.

One can use the *t* test for two means or its range analog (*M* test).

Example 9.3 *t Test*

Step 1. Calculate the combined standard deviation: The quantity s^2 is the sum squares for each analyst divided by degrees of freedom for each analyst.

$$s^2 = (0.0214 + 0.0295)/(5 + 3)$$

$$= 0.00636,$$

$$s = 0.0797.$$

$$t = \frac{59.62 - 59.15}{0.0797} \left[\frac{(6 \times 4)}{(6 + 4)} \right]^{1/2}$$

$$= 9.13.$$

There are 8 *df*. In Table XI, *t* for $\alpha = 0.01$, 8 *df* = 3.36. The difference between analysts is significant.

Example 9.4 *M Test*

$$M = (\bar{X}_1 - \bar{X}_2)/(R_1 + R_2)$$

$$= (59.62 - 59.15)/(0.24 + 0.18)$$

$$= 1.119.$$

Critical values for *M* are given in Table IV.
For $n_1 = 4$ and $n_2 = 6$, $M = 1.119$ is highly significant.

9.5 ACCURACY

The intuitive approach might lead one to believe that both assays are accurate within the limits of experimental error since pure ephedrine hydrochloride has a theoretical carbon content of 59.55%.

The relative error for Analyst H is 4 parts in 600 and for FWP about 7 parts in 6000.

The statistical approach would be to test by the t test or its analog the L test.

Example 9.5 *t Test*

For Analyst FWP:

Step 1.   Calculate s as done in Example 9.1

$$s = 0.099.$$

Step 2.   Substitute in

$$t = \frac{\bar{X} - \mu}{s} \sqrt{n}$$

$$= \frac{59.62 - 59.55}{0.099} \sqrt{4}$$

$$= 1.41.$$

Similarly for Analyst H

$$t = \frac{59.55 - 59.15}{0.065} \sqrt{6}$$

$$= 15.1.$$

Consulting Table XI, for 3 df, $t = 5.84$ and for 5 df $t = 4.03$ at a probability level of 0.01. Hence it is evident that the results of Analyst H are significantly low.

Example 9.6 *L Test*

Substitute in

$$L = \frac{\bar{X} - \mu}{R}.$$

For Analyst H,

$$L = \frac{59.55 - 59.15}{0.18} = 2.22.$$

For Analyst FWP,

$$L = \frac{59.62 - 59.55}{0.24} = 0.29.$$

Critical values for L are given in Table III. Consulting this table, it is again evident that the results of Analyst H are significantly low.

9.6 PRECISION OF OPTICAL ROTATION MEASUREMENTS

When reading the optical rotation of a compound, the analyst makes several blank readings with the solvent to establish a zero point, and then makes several readings of the optically active solution, usually approaching the match point from different sides.

The experimental design is the simple one of two averages with their experimental errors.

Example 9.7

The readings given in the following tabulation were obtained by an analyst:

Sample	Zero
35.884	0.004
35.882	0.004
35.882	0.002
35.883	0.002
35.885	0.004
35.884	0.002
$\bar{X} = 35.8833$	$\bar{X} = 0.0030$
$R = 0.003$	$R = 0.002$
$\Sigma R = 0.005$	

In Table I, at $k = 2$, $n = 6$: $A = 0.125$. The 95% confidence limits are:

$$CL = A(\Sigma R)$$

$$= 0.005(0.125) = 0.0006.$$

Therefore, the angular rotation of the compound is:

$$(35.8833 - 0.0030) \pm 0.0006 = 35.8797\text{--}35.8809$$

and the specific rotation is:

$$\frac{35.8803}{1d} \pm \frac{0.0006}{1d}.$$

9.7 PRECISION OF COLORIMETRIC ANALYSIS

It is common practice in colorimetry to set the blank at zero and run a series of replicated standards and unknowns at about the same concentration. This assumes equal slopes and therefore:

$$X_u = X_s \bar{U}/\bar{Y},$$

where $\bar{U}$ and $\bar{Y}$ are the average of the readings of the unknown and the standard, respectively.

Example 9.8

The results of a colorimetric analysis are given in the following tabulation. The blank absorbence is zero. The value for X_s is 1.00.

Calculations:

$$X_u = 1.00 \times (47.34/43.86) = 1.079.$$

Slope of standard line b equals $43.86/1.00 = 43.86$.

Sum of range of X equals $(0.5 + 0.6)/43.86 = 0.025$.

$$95\% \; CL = \bar{X} \pm A(\Sigma R)$$

$$= 1.079 \pm 0.155(0.025)$$

$$= 1.075 \text{ to } 1.083.$$

Absorbence

Y	U
43.7	47.0
43.9	47.1
43.6	47.6
44.0	47.5
44.1	47.5
$\bar{Y} = 43.86$	$\bar{U} = 47.34$
Range: 0.5	Range: 0.6

9.8 REDUCED SAMPLE SIZE

The question, "What size sample shall I take?" was answered in Section 7.5, where equations were given to calculate the number of replications necessary for a given set of risks and specifications. However, there are times when circumstances limit the chemist to a smaller number of analyses. If he knows the standard deviation of the analytical error, he can use a reduced sample size by increasing the risks, and the increased risk can be calculated with the help of Table 9.2.

Table 9.2 is set up on the assumption that the risks of accepting a bad sample and of rejecting a good sample are equal.

Example 9.9 *Reduced Sample, Single Limit*

In Example 8.3 it was postulated that the specification limit was 95%, that the known standard deviation was 3.1%, and that we wanted 95% assurance that the product would meet these specifications. Example 8.3 called for making five replications. Suppose it is

Table 9.2

Probability of error	One-tailed (T_1)	Two-tailed (T_2)	Assurance (%)
0.40	0.51	1.10	60
0.20	1.68	2.12	80
0.10	2.56	2.92	90
0.08	2.80	3.15	92
0.06	3.11	3.44	94
0.05	3.29	3.60	95
0.04	3.50	3.80	96
0.03	3.76	4.05	97
0.02	4.10	4.37	98
0.01	4.66	4.81	99
0.005	5.15	—	99.5
0.001	6.18	—	99.9

desired to reduce the number of analyses to one or two. What risks are involved?

Step 1. Substitute in Eq. (9.2) and solve for T

$$T = \frac{d\sqrt{n}}{s}. \tag{9.2}$$

For $n = 1$,

$$T = \frac{5}{3.1} = 1.61.$$

This is T_1 (one-tailed) in Table 9.2.

The chemist has lowered his assurance from 95% to about 80%.

For $n = 2$,

$$T = \frac{1.41(5)}{3.1} = 2.27.$$

Using the average of two analyses will give about 90% assurance.

For $n = 3$,

$$T = 2.79.$$

Three analyses will give about 92% assurance.

The analyst can reduce the sample size to two or three analyses with very little increased risk.

Example 9.10 *Reduced Sample, Double Limit*

In Example 8.4 we postulated a $\pm 5\%$ specification with a standard deviation of 3.1, and 95% assurance. To investigate the risks of decreasing the sample size we substitute in Eq. (9.2). Using T_2 (two-tailed) in Table 9.2

For $n = 1$,

$$T = 1.61 \qquad \text{(assurance is less than 80\%).}$$

For $n = 2$,

$$T = 2.27 \qquad \text{(assurance is better than 80\%).}$$

For $n = 3$,

$$T = 2.79 \qquad \text{(assurance is about 80–90\%).}$$

For $n = 4$,

$$T = 3.22 \qquad \text{(assurance is 92\%).}$$

If the chemist is willing to accept 80–90% assurance of not making an error, he should make three analyses, average the results and accept if the average falls within the limits set by

$$d = 1.28s/\sqrt{n}; \qquad (9.3)$$

$$d = 1.28(3.1)/1.732;$$

$$d = 2.29.$$

Limits $100 \pm 2.29 = 97.71–102.29\%$.

9.9 COMPLIANCE WITH SPECIFICATIONS

When a substance is at or near a specification limit, the variation due to experimental error will often produce an analytical result which is outside of specifications. For example, if the product is at the upper specification limit, 50% of the analyses will show it to be too high. If it is 1 standard deviation below the upper specification, about 23% of the analyses will show it to be high, and if it is 2 standard deviations below the limit, 5% of the analyses will reject the product. The converse is also true: A product 1 standard deviation above specifications will be accepted 23% of the time. In these borderline cases the analyst has the problem of deciding if the product is acceptable.

Example 9.11

The specified upper limit of impurity in a substance is 0.01%. The analyst obtains a result of 0.014%. He repeats the analysis twice and gets 0.009% and 0.012%. The average of the three analyses is 0.0113%. On the basis of the experimental error, is the lot acceptable? This is equivalent to asking if there is a significant difference between $\bar{X} = 0.0113\%$ and $\bar{X}' = 0.010\%$, and we use a one-tailed L test (Section 4.2).

$$L = (0.0113 - 0.0100)/(0.013 - 0.009)$$

$$= 0.325.$$

From Table III, $L_{0.05}$ for a one-tailed test is 0.88, and hence the analyst may accept the batch.

This philosophy of compliance with the specification guarantees the analyst that he will reject a lot that complies with the specification only 5% of the time. However, he is taking a larger risk of accepting material that does not meet specifications: In this example, he will accept rejectable material about 25% of the time.

A more conservative plan would be to accept the product only if the analysis indicates the impurity is no more than the specification limit. This plan will not eliminate the acceptance of defective material, but it will minimize it. For example, if the analytical error of the test in this example is known to be $s = 0.001$, this plan will accept material which contains up to

$$0.01 + 3.29(0.001) = 0.013\%$$

5% of the time.

If it is necessary that no material containing more than 0.01% impurity be accepted, the limit should be set at

$$0.01 - 3.29(0.001) = 0.007\%.$$

The factor, 3.29, is the T_1 factor for 95% assurance from Table 9.2.

9.10 CONTROL CHARTS IN THE ANALYTICAL LABORATORY

The average–range control chart is a very useful tool in the control laboratory where data are being generated rapidly. The technique was illustrated in Chapter 6. Tables IXD, E, and F give 95 and 99% probability limits for $A_2{}^*$, $D_3{}^*$, and $D_4{}^*$.

A point out of control means either a poor analysis or a product that differs significantly from the normal. The analyst must use his professional judgment to decide which; the statistical technique cannot do this for him. Although keeping control charts may seem to be time consuming, the visual representation makes it possible to see trends and thus to anticipate trouble. For example, an upward trend on a control chart could indicate a standard solution is losing strength.

9.11 INTERLABORATORY STUDIES

9.11.1 *Purpose of Study*

An interlaboratory study is used to determine one of the following:

(1) The accuracy and precision of an analytical method in several laboratories.
(2) A comparison of two methods of analysis in several laboratories.
(3) To establish a standard value for a product or a measurement.

A well-designed experiment will give as much information about the laboratory as about the product. This point will be brought out in further discussion.

9.11.2 *Design of the Experiment*

The experiment should be designed so that the results are objective. This means it should yield data that will give some basic measurement of the precision which can be expected within a single laboratory. With this measurement as a basis, objective tests can be made on the variation between laboratories, or between test methods.

The design should be kept simple. If too many methods or too many samples are investigated, the study becomes burdensome to the laboratory, and the design becomes complicated in its interpretation.

As it happens all three of the purposes of a collaborative experiment can be interpreted by the same type of statistical analysis.

9.11.3 *Analysis of the Data*

The analysis of the data should be presented in a form that is easy to understand. Too often the jargon of the statistician means little or nothing to the chemist. This is a problem in communications. Statements such as "There is a significant difference between laboratories," and "The laboratory–method interaction is significant at the 0.05 probability level" may be perfectly true, but are too general to be useful. A chemist reads the statistical analysis, says, "Too bad those other laboratories are out in left field," and files the report. We must know which laboratories are out of line before we know why their results do not jibe.

Dr. W. J. Youden has published several articles describing a design for interlaboratory studies and a method for analyzing the results. The analysis is graphical, and therefore easy to interpret. As described below the design is for the purpose of testing the precision of a method [Purpose (1)], but it can be adapted to the other purposes of interlaboratory studies.

9.11.4 *Youden's Method*

(1) Two different samples are sent to the collaborating laboratories. It is preferable, but not necessary, that the samples be similar.

(2) The laboratory runs one test on each sample, giving a pair of results, say X and Y.

(3) A graph is prepared in which the x axis covers the range of Sample X, and the y axis the range of values of Sample Y.

(4) Each pair of results reported by a laboratory is used to plot a point on the graph thereby giving a scatter diagram. Points which are obviously in error can be detected immediately and discarded from the subsequent calculations.

(5) The averages of X and Y are calculated as well as the range between X and Y for each laboratory.

(6) A 45° line running from the lower left quadrant to the upper right quadrant through the point $\bar{X}\bar{Y}$ is drawn.

(7) Interpretation of the graph is based upon the distribution of the points on the graph.

If there were no errors in the method, all the laboratories would obtain the same results and the points would all be at $\bar{X}\bar{Y}$. Hence any deviation from the grand average is caused by analytical error. These are of three kinds:

(*a*) *Erratic errors:* a high result for one method and a low result for the other.

(*b*) *Analytical variability:* the usual indeterminate laboratory error.

(*c*) *Laboratory bias:* the tendency of a laboratory always to obtain low (or high) results.

Erratic errors magnify the within-laboratory variability. They are easily found from their position on the graph and thrown out at Step (4).

If the error is only analytical variability, the points will be clustered around $\bar{X}\bar{Y}$, with a uniform distribution in all four quadrants.

If the only error were laboratory bias, the points would all lie on a 45° line.

In practice, the errors are often a combination of both analytical variability and laboratory bias, so that the points may tend toward an elliptical pattern along the 45° line.

Example 9.12

Twelve laboratories collaborated in the investigation of two analytical methods, X and Y, with the results listed in Columns X and Y of Table 9.3.

Procedure:

Step 1.   Plot the 12 points derived from the 24 analyses submitted by the collaborators.

Step 2.   Examine the scatter diagram and discard the points obviously in error. Laboratory 11 has a biased result: It shows good agreement, but both tests are obviously low. Laboratory 12 has made an error with Method X. Both laboratories' results are therefore discarded.

Step 3.   Calculate $\bar{X} = 99.97$ and $\bar{Y} = 100.02$.

Step 4.   Draw a 45° line through $\bar{X}\bar{Y}$.

Step 5.   Calculate $T = (X + Y)$ for each laboratory.

Table 9.3

Laboratory	Method X	Method Y	$D = X - Y$	$T = X + Y$
1	100.0	99.5	0.5	199.5
2	99.7	99.7	0.0	199.4
3	99.4	101.0	−1.6	200.4
4	101.1	100.4	0.7	201.5
5	99.5	99.8	−0.3	199.3
6	99.8	100.7	−0.9	200.5
7	100.2	99.2	1.0	199.4
8	99.4	99.7	−0.3	199.1
9	100.8	100.7	0.1	201.5
10	99.8	99.5	0.3	199.3
11	97.9	98.1	—	—
12	97.5	98.8	—	—
Averages (10)	99.97	100.02	−0.05	199.99

Step 6. Calculate

$$s_T^2 = \Sigma(T - \bar{T})^2/2(n - 1)$$
$$= 7.6690/2(9) = 0.4260,$$
$$s_T = 0.65.$$

The variance, s_T^2, is the "between laboratories" variance one would get from doing an analysis of variance on these data.

Step 7. Calculate $D = (X - Y)$ for each laboratory.

Step 8. Calculate

$$s_D^2 = \Sigma(D - \bar{D})^2/2(n - 1)$$
$$= 5.3650/2(9) = 0.2980,$$
$$s_D = 0.55.$$

The variance, s_D^2, is the residual (random) variance one would get from doing an analysis of variance on these data. s_D is the standard deviation of the random error.

Step 9. Test for significance

$$F = 0.4260/0.2980 = 1.43.$$

From Table VIIIA the critical value of F at $f_1 = 8$ and $f_2 = 8$ is 3.44.

Step 10.

$$s_T^2 = 2s_B^2 + s_D^2$$
$$s_B^2 = (0.4260 - 0.2980)/2 = 0.0640$$
$$s_B = 0.25,$$

where s_B is the standard deviation of the bias of the data.

Step 11. Calculate the circular confidence limits:

95% $CL = 2.45(s)$,
99% $CL = 3.04(s)$.

In this example we shall use 95% CL:

$$2.45(0.55) = 1.35.$$

Step 12. Draw a circle with center $\bar{X} = 99.97$, $\bar{Y} = 100.02$, and radius 1.35. Note all laboratory results except 11 and 12 are inside the circle, hence the only between laboratory difference is due to random error.

Interpretation of study:

(1) Ten of the twelve collaborating laboratories produced valid results.

(2) There is no significant difference between methods; the $\bar{X}\bar{Y}$ point is 99.97, 100.02. [If there were a bias of say -0.5%

in the X method, it is obvious that $\bar{X}\bar{Y}$ would be 99.47, 100.02 (see Fig. 9.1).]

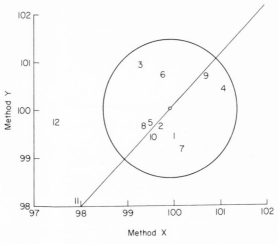

Figure 9.1.

REFERENCE

Youden, W. J. (1967). "Statistical Techniques for Collaborative Tests." The Association of Official Analytical Chemists, Washington, D.C.

Appendix

Table I $CI = \bar{X} \pm A(\Sigma R)$. Confidence Interval for Averages [a]

						n				
Prob.	k	2	3	4	5	6	7	8	9	10
0.05	1	6.36	1.30	0.719	0.505	0.402	0.336	0.291	0.256	0.232
0.01		31.9	3.00	1.36	0.865	0.673	0.514	0.430	0.379	0.338
0.05	2	0.879	0.316	0.206	0.154	0.125	0.106	0.093	0.084	0.076
0.01		2.11	0.474	0.312	0.227	0.179	0.150	0.131	0.116	0.105
0.05	3	0.360	0.156	0.104	0.079	0.065	0.056	0.049	0.044	0.040
0.01		0.660	0.273	0.150	0.112	0.091	0.077	0.068	0.060	0.054
0.05	4	0.210	0.096	0.065	0.050	0.042	0.036	0.032	0.028	0.026
0.01		0.350	0.142	0.092	0.070	0.057	0.048	0.043	0.038	0.035
0.05	5	0.140	0.066	0.046	0.035	0.030	0.025	0.022	0.020	0.018
0.01		0.226	0.095	0.063	0.049	0.040	0.034	0.030	0.027	0.025
0.05	6	0.102	0.050	0.034	0.027	0.022	0.019	0.017	0.015	0.014
0.01		0.157	0.070	0.047	0.036	0.030	0.026	0.023	0.020	0.019
0.05	7	0.079	0.039	0.027	0.021	0.018	0.015	0.013	0.012	0.011
0.01		0.117	0.055	0.037	0.029	0.024	0.020	0.018	0.016	0.015
0.05	8	0.063	0.032	0.022	0.017	0.014	0.012	0.011	0.010	0.009
0.01		0.094	0.044	0.030	0.023	0.019	0.016	0.014	0.013	0.012
0.05	9	0.053	0.027	0.018	0.014	0.012	0.010	0.009	0.008	0.007
0.01		0.076	0.036	0.025	0.019	0.016	0.014	0.012	0.011	0.010
0.05	10	0.044	0.023	0.016	0.012	0.010	0.009	0.008	0.007	0.006
0.01		0.064	0.031	0.021	0.016	0.014	0.012	0.010	0.009	0.008

[a] Given k subgroups of n numbers, the confidence interval is $\bar{X} \pm A(\Sigma R)$.

$$A = \frac{t}{kc_1} (\sqrt{kn})$$

Table II $TI = \bar{X} \pm I(\Sigma R)$. *Tolerance Interval for Individuals*

Prob.	k	n								
		2	3	4	5	6	7	8	9	10
0.05	1	8.99	2.25	1.44	1.13	0.985	0.889	0.823	0.768	0.734
0.01		45.1	5.20	2.72	1.93	1.65	1.36	1.22	1.14	1.07
0.05	2	1.76	0.774	0.583	0.487	0.433	0.397	0.372	0.356	0.340
0.01		4.22	1.16	0.882	0.718	0.620	0.561	0.524	0.492	0.470
0.05	3	0.882	0.486	0.360	0.306	0.276	0.257	0.240	0.229	0.219
0.01		1.62	0.819	0.520	0.434	0.386	0.353	0.333	0.312	0.296
0.05	4	0.594	0.332	0.260	0.224	0.206	0.190	0.181	0.168	0.164
0.01		0.990	0.492	0.368	0.313	0.279	0.254	0.243	0.228	0.221
0.05	5	0.443	0.256	0.206	0.175	0.164	0.148	0.139	0.134	0.127
0.01		0.715	0.368	0.282	0.245	0.219	0.201	0.190	0.181	0.177
0.05	6	0.353	0.212	0.166	0.148	0.132	0.123	0.118	0.110	0.108
0.01		0.544	0.297	0.230	0.197	0.180	0.168	0.159	0.147	0.145
0.05	7	0.296	0.179	0.143	0.124	0.117	0.105	0.097	0.095	0.092
0.01		0.438	0.252	0.196	0.172	0.156	0.140	0.135	0.127	0.124
0.05	8	0.252	0.157	0.124	0.108	0.097	0.090	0.088	0.085	0.080
0.01		0.376	0.216	0.170	0.145	0.132	0.120	0.112	0.110	0.108
0.05	9	0.225	0.140	0.108	0.094	0.088	0.079	0.076	0.072	0.066
0.01		0.322	0.187	0.150	0.127	0.118	0.111	0.101	0.099	0.095
0.05	10	0.197	0.126	0.101	0.085	0.077	0.075	0.072	0.066	0.060
0.01		0.286	0.170	0.133	0.113	0.108	0.100	0.089	0.085	0.080

Table III   Values of L That Will Be Exceeded with a Probability P^a

	$L = (\bar{X}_1 - \bar{X}_2)/R$			
	P_1			
n	0.05	0.025	0.01	0.005
2	3.16	6.36	15.9	31.9
3	0.883	1.30	2.10	3.00
4	0.533	0.719	1.07	1.36
5	0.390	0.505	0.692	0.865
6	0.313	0.402	0.529	0.673
7	0.264	0.336	0.433	0.514
8	0.229	0.291	0.369	0.430
9	0.205	0.256	0.324	0.379
10	0.185	0.232	0.290	0.338
	0.1	0.05	0.02	0.01
	P_2			

[a] P_1 is the α for a single-tail test; P_2 is the α for a two-tail test.

Table IV Values of M That Will Be Exceeded with a
Probability P

$$M = \frac{\bar{X}_1 - \bar{X}_2}{R_1 + R_2}$$

		P_1			
		0.05	0.25	0.01	0.005
		P_2			
n_1	n_2	0.10	0.05	0.02	0.01
2	2	1.161	1.714	2.776	3.958
	3	0.693	0.915	1.255	1.557
	4	0.556	0.732	1.002	1.242
	5	0.478	0.619	0.827	1.008
	6	0.429	0.549	0.721	0.865
	7	0.396	0.502	0.652	0.776
	8	0.372	0.469	0.603	0.713
	9	0.353	0.443	0.567	0.666
	10	0.338	0.423	0.538	0.630
	15	0.294	0.363	0.455	0.526
	20	0.270	0.333 .	0.414	0.475
3	3	0.487	0.635	0.860	1.050
	4	0.398	0.511	0.663	0.814
	5	0.339	0.429	0.556	0.660
	6	0.311	0.391	0.501	0.590
	7	0.288	0.360	0.458	0.536
	8	0.271	0.338	0.427	0.498
	9	0.258	0.321	0.404	0.469
	10	0.248	0.307	0.385	0.446
	15	0.216	0.266	0.330	0.378
	20	0.200	0.245	0.302	0.344
4	4	0.322	0.407	0.526	0.620
	5	0.282	0.353	0.450	0.528
	6	0.256	0.319	0.403	0.469
	7	0.237	0.294	0.370	0.429
	8	0.224	0.276	0.346	0.399
	9	0.213	0.263	0.327	0.377
	10	0.204	0.252	0.313	0.359
	15	0.178	0.218	0.268	0.306
	20	0.164	0.200	0.246	0.279

Table IV—Continued

$$M = \frac{\bar{X}_1 - \bar{X}_2}{R_1 + R_2}$$

		P_1			
		0.05	0.25	0.01	0.005
		P_2			
n_1	n_2	0.10	0.05	0.02	0.01
5	5	0.247	0.307	0.387	0.450
	6	0.224	0.277	0.347	0.402
	7	0.208	0.256	0.319	0.368
	8	0.195	0.240	0.299	0.343
	9	0.186	0.228	0.282	0.323
	10	0.178	0.218	0.270	0.309
	15	0.155	0.189	0.232	0.263
	20	0.142	0.173	0.212	0.240
6	6	0.203	0.250	0.312	0.359
	7	0.188	0.240	0.287	0.329
	8	0.177	0.217	0.268	0.307
	9	0.168	0.206	0.254	0.289
	10	0.161	0.197	0.242	0.276
	15	0.139	0.169	0.207	0.235
	20	0.128	0.155	0.189	0.214
7	7	0.174	0.213	0.263	0.301
	8	0.163	0.200	0.246	0.281
	9	0.155	0.189	0.233	0.265
	10	0.148	0.181	0.222	0.252
	15	0.128	0.155	0.189	0.214
	20	0.117	0.142	0.172	0.195
8	8	0.153	0.187	0.231	0.262
	9	0.145	0.177	0.217	0.247
	10	0.139	0.169	0.207	0.235
	15	0.119	0.144	0.176	0.199
	20	0.109	0.132	0.160	0.180
9	9	0.137	0.167	0.205	0.233
	10	0.131	0.160	0.195	0.221
	15	0.112	0.136	0.165	0.187
	20	0.102	0.124	0.150	0.169

Table IV—Continued

$$M = \frac{\bar{X}_1 - \bar{X}_2}{R_1 + R_2}$$

		P_1			
		0.05	0.25	0.01	0.005
				P_2	
n_1	n_2	0.10	0.05	0.02	0.01
10	10	0.125	0.152	0.186	0.210
	12	0.116	0.141	0.171	0.194
	14	0.109	0.133	0.161	0.182
	16	0.104	0.126	0.153	0.173
	18	0.100	0.121	0.147	0.165
	20	0.097	0.117	0.142	0.160
12	12	0.107	0.130	0.158	0.178
	14	0.101	0.122	0.148	0.167
	16	0.096	0.116	0.140	0.158
	18	0.092	0.111	0.134	0.151
	20	0.089	0.107	0.130	0.146
14	14	0.094	0.114	0.138	0.156
	16	0.090	0.108	0.131	0.147
	18	0.086	0.104	0.125	0.141
	20	0.083	0.101	0.121	0.135
16	16	0.085	0.103	0.124	0.139
	18	0.081	0.098	0.118	0.133
	20	0.078	0.094	0.114	0.128
18	18	0.077	0.093	0.113	0.126
	20	0.074	0.090	0.108	0.121
20	20	0.071	0.086	0.104	0.116

[a] Condensed from Table I, from Moore, P. G., *Biometrika*, 1957, **44**, p. 487. This table is reproduced by permission from *Biometrika*.

Table V Single Classification Factor (c_1) to Estimate **Standard** *Deviation from Range, and Equivalent Degrees of Freedom (f)[a]*

$$s = \frac{\bar{R}}{c_1}$$

												n							
	2		3		4		5		6		7		8		9		10	
k	f	c_1	f	c_1	f	c_1	f	c_1	f	c_1	f	c_1	f	c_1	f	c_1	f	c_1
1	1.0	1.41	2.0	1.91	2.9	2.24	3.8	2.48	4.7	2.67	5.5	2.83	6.3	2.96	7.0	3.08	7.7	3.18
2	1.9	1.28	3.8	1.81	5.7	2.15	7.5	2.40	9.2	2.60	10.8	2.77	12.3	2.91	13.8	3.02	15.1	3.13
3	2.8	1.23	5.7	1.77	8.4	2.12	11.1	2.38	13.6	2.58	16.0	2.75	18.3	2.89	20.5	3.01	22.6	3.11
4	3.7	1.21	7.5	1.75	11.2	2.11	14.7	2.37	18.1	2.57	21.3	2.74	24.4	2.88	27.3	3.00	30.1	3.10
5	4.6	1.19	9.3	1.74	13.9	2.10	18.4	2.36	22.6	2.56	26.6	2.73	30.4	2.87	34.0	2.99	37.5	3.10
6	5.5	1.18	11.1	1.73	16.6	2.09	22.0	2.36	27.1	2.56	31.9	2.73	36.4	2.87	40.8	2.99	45.0	3.10
7	6.4	1.17	12.9	1.72	19.4	2.08	25.6	2.35	31.5	2.56	37.1	2.73	42.5	2.87	47.6	2.98	52.4	3.09
8	7.2	1.16	14.8	1.71	22.1	2.08	29.3	2.35	36.0	2.55	42.4	2.72	48.5	2.86	54.3	2.98	59.8	3.09
9	8.1	1.15	16.6	1.70	24.9	2.07	32.9	2.34	40.5	2.55	47.7	2.72	54.5	2.86	61.1	2.98	67.3	3.09
10	9.0	1.14	18.4	1.69	27.6	2.07	36.5	2.34	44.9	2.55	52.9	2.72	60.6	2.86	67.8	2.98	74.8	3.09
d_n		1.13		1.69		2.06		2.33		2.53		2.70		2.85		2.97		3.08
CD	0.88		1.82		2.74		3.62		4.47		5.27		6.03		6.76		7.45	

[a] This table is reproduced with the permission of Professor E. S. Pearson from David, H. A., "Further applications of range to analysis of variance." *Biometrika,* **38,** p. 393, 1951.

Table VIA Double Classification Factor (c_2) to Estimate Standard Deviation from Range, and Equivalent Degrees of Freedom (f)[a]

$$s = \frac{\bar{R}}{c_2}$$

	n																
	2		3		4		5		6		7		8		9	
k	f	c_2	f	c_2	f	c_2	f	c_2	f	c_2	f	c_2	f	c_2	f	c_2
2	1.0	1.0	2.0	1.35	2.9	1.58	3.8	1.75	4.7	1.89	5.5	2.00	6.3	2.10	7.0	2.18
3	1.9	1.05	3.7	1.48	5.6	1.76	7.4	1.96	9.3	2.12	11.3	2.26	13.4	2.37	15.7	2.46
4	2.7	1.07	5.4	1.54	8.2	1.84	11.0	2.06	13.9	2.23	16.9	2.38	20.1	2.50	23.6	2.60
5	3.6	1.08	7.2	1.57	10.9	1.88	14.6	2.12	18.5	2.30	22.4	2.45	26.6	2.57	31.1	2.68
6	4.5	1.09	8.9	1.59	13.6	1.91	18.2	2.15	23.0	2.34	27.9	2.49	33.0	2.62	38.3	2.73
7	5.4	1.09	10.7	1.61	16.3	1.93	21.8	2.18	27.6	2.37	33.3	2.52	39.3	2.65	45.4	2.76
8	6.3	1.10	12.5	1.62	19.0	1.95	25.4	2.20	32.1	2.39	38.7	2.55	45.6	2.68	52.5	2.79
9	7.1	1.10	14.3	1.63	21.7	1.96	29.0	2.21	36.6	2.41	44.0	2.57	51.8	2.70	59.6	2.81
10	8.1	1.10	16.1	1.63	24.4	1.97	32.6	2.22	41.0	2.42	49.3	2.58	57.9	2.71	66.6	2.83
20	16.7	1.11	33.9	1.66	51.5	2.02	68.8	2.28	86.0	2.48	103.0	2.64	119.0	2.78	134.0	2.90
CD	0.87		1.80		2.71		3.62		4.50		5.33		6.10		6.79	
d_n		1.13		1.69		2.06		2.33		2.53		2.70		2.85		2.97

[a] This table is reproduced with the permission of Professor E. S. Pearson from David, H. A., "Further applications of range to analysis of variance." *Biometrika*, **38**, p. 393, 1951.

Table VIB. *Triple Classification Factor (c_3) to Estimate Standard Deviation from Range, and Equivalent Degrees of Freedom (f')*

$$s = \frac{\overline{R}}{c_3}$$

Example: $k = 3, n = 3, m = 3$
$f' = mf = 3.6$

$f = \frac{3.6}{3} = 1.2$

n

k	2 f'	2 c_3	3 f'	3 c_3	4 f'	4 c_3	5 f'	5 c_3	6 f'	6 c_3	7 f'	7 c_3	8 f'	8 c_3	9 f'	9 c_3
2	0.9	0.84	1.9	1.23	2.8	1.48	3.7	1.67	4.6	1.81	5.4	1.94	6.1	2.04	6.9	2.13
3	1.8	0.95	3.6	1.40	5.6	1.70	7.2	1.92	9.2	2.08	11.1	2.24	13.2	2.35	15.7	2.45
4	2.6	1.00	5.3	1.48	8.1	1.80	10.9	2.04	13.8	2.22	16.8	2.36	19.9	2.48	23.5	2.59
5	3.5	1.03	7.0	1.53	10.7	1.86	14.5	2.10	18.3	2.28	22.3	2.43	26.5	2.57	30.9	2.67
6	4.4	1.05	8.8	1.56	13.4	1.90	18.1	2.14	22.9	2.33	27.7	2.48	32.8	2.61	38.1	2.72
7	5.2	1.06	10.6	1.59	16.2	1.92	21.7	2.17	27.5	2.36	33.2	2.51	39.1	2.64	45.2	2.76
8	6.1	1.07	12.3	1.60	18.8	1.94	25.2	2.19	32.1	2.38	38.6	2.54	45.3	2.67	52.2	2.79
9	7.0	1.08	14.1	1.61	21.6	1.95	28.8	2.20	36.6	2.40	44.0	2.56	51.6	2.69	59.2	2.81
10	7.9	1.09	15.9	1.62	24.3	1.96	32.5	2.21	41.1	2.41	49.3	2.57	57.7	2.71	66.2	2.82

a This table is reproduced with the permission of Professor E. S. Pearson from David, H. A., "Further applications of range to analysis of variance." *Biometrika*, **38**, p. 393, 1951.

Table VIIA Upper 5% Points of the Studentized Range[a]

$$q = \frac{X_n - X_1}{s_x}$$

k

f	2	3	4	5	6	7	8	9	10	11	12	13	14	15	16	17	18	19	20
1	18.0	27.0	32.8	37.1	40.4	43.1	45.4	47.4	49.1	50.6	52.0	53.2	54.3	55.4	56.3	57.2	58.0	58.8	59.6
2	6.09	8.3	9.9	10.9	11.7	12.4	13.0	13.5	14.0	14.4	14.7	15.1	15.4	15.7	15.9	16.1	16.4	16.6	16.8
3	4.50	5.91	6.82	7.50	8.04	8.48	8.85	9.18	9.46	9.72	9.95	10.15	10.35	10.52	10.69	10.84	10.98	11.11	11.24
4	3.93	5.04	5.76	6.29	6.71	7.05	7.35	7.60	7.83	8.03	8.21	8.37	8.52	8.66	8.79	8.91	9.03	9.13	9.23
5	3.64	4.60	5.22	5.67	6.03	6.33	6.58	6.80	6.99	7.17	7.32	7.47	7.60	7.72	7.83	7.93	8.03	8.12	8.21
6	3.46	4.34	4.90	5.31	5.63	5.89	6.12	6.32	6.49	6.65	6.79	6.92	7.03	7.14	7.24	7.34	7.43	7.51	7.59
7	3.34	4.16	4.68	5.06	5.36	5.61	5.82	6.00	6.16	6.30	6.43	6.55	6.66	6.76	6.85	6.94	7.02	7.09	7.17
8	3.26	4.04	4.53	4.89	5.17	5.40	5.60	5.77	5.92	6.05	6.18	6.29	6.39	6.48	6.57	6.65	6.73	6.80	6.87
9	3.20	3.95	4.42	4.76	5.02	5.24	5.43	5.60	5.74	5.87	5.98	6.09	6.19	6.28	6.36	6.44	6.51	6.58	6.64
10	3.15	3.88	4.33	4.65	4.91	5.12	5.30	5.46	5.60	5.72	5.83	5.93	6.03	6.11	6.20	6.27	6.34	6.40	6.47
11	3.11	3.82	4.26	4.57	4.82	5.03	5.20	5.35	5.49	5.61	5.71	5.81	5.90	5.99	6.06	6.14	6.20	6.26	6.33
12	3.08	3.77	4.20	4.51	4.75	4.95	5.12	5.27	5.40	5.51	5.62	5.71	5.80	5.88	5.95	6.03	6.09	6.15	6.21
13	3.06	3.73	4.15	4.45	4.69	4.88	5.05	5.19	5.32	5.43	5.53	5.63	5.71	5.79	5.86	5.93	6.00	6.05	6.11
14	3.03	3.70	4.11	4.41	4.64	4.83	4.99	5.13	5.25	5.36	5.46	5.55	5.64	5.72	5.79	5.85	5.92	5.97	6.03

15	3.01	3.67	4.08	4.37	4.60	4.78	4.94	5.08	5.20	5.31	5.40	5.49	5.58	5.65	5.72	5.79	5.85	5.90	5.96
16	3.00	3.65	4.05	4.33	4.56	4.74	4.90	5.03	5.15	5.26	5.35	5.44	5.52	5.59	5.66	5.72	5.79	5.84	5.90
17	2.98	3.63	4.02	4.30	4.52	4.71	4.86	4.99	5.11	5.21	5.31	5.39	5.47	5.55	5.61	5.68	5.74	5.79	5.84
18	2.97	3.61	4.00	4.28	4.49	4.67	4.82	4.96	5.07	5.17	5.27	5.35	5.43	5.50	5.57	5.63	5.69	5.74	5.79
19	2.96	3.59	3.98	4.25	4.47	4.65	4.79	4.92	5.04	5.14	5.23	5.32	5.39	5.46	5.53	5.59	5.65	5.70	5.75
20	2.95	3.58	3.96	4.23	4.45	4.62	4.77	4.90	5.01	5.11	5.20	5.28	5.36	5.43	5.49	5.55	5.61	5.66	5.71
24	2.92	3.53	3.90	4.17	4.37	4.54	4.68	4.81	4.92	5.01	5.10	5.18	5.25	5.32	5.38	5.44	5.50	5.54	5.59
30	2.89	3.49	3.84	4.10	4.30	4.46	4.60	4.72	4.83	4.92	5.00	5.08	5.15	5.21	5.27	5.33	5.38	5.43	5.48
40	2.86	3.44	3.79	4.04	4.23	4.39	4.52	4.63	4.74	4.82	4.91	4.98	5.05	5.11	5.16	5.22	5.27	5.31	5.36
60	2.83	3.40	3.74	3.98	4.16	4.31	4.44	4.55	4.65	4.73	4.81	4.88	4.94	5.00	5.06	5.11	5.16	5.20	5.24
120	2.80	3.36	3.69	3.92	4.10	4.24	4.36	4.48	4.56	4.64	4.72	4.78	4.84	4.90	4.95	5.00	5.05	5.09	5.13
∞	2.77	3.31	3.63	3.86	4.03	4.17	4.29	4.39	4.47	4.55	4.62	4.68	4.74	4.80	4.85	4.89	4.93	4.97	5.01

[a] This table is reproduced with the permission of Professor E. S. Pearson from May, J. M., "Extended and corrected tables of the upper percentage points of the studentized range." *Biometrika*, **39**, p. 192, 1952.

Table VIIB   Upper 1% Points of the Studentized Range[a]

$$q = \frac{X_n - X_1}{s_x}$$

k

f	2	3	4	5	6	7	8	9	10	11	12	13	14	15	16	17	18	19	20
1	90.0	135	164	186	202	216	227	237	246	253	260	266	272	277	282	286	290	294	298
2	14.0	19.0	22.3	24.7	26.6	28.2	29.5	30.7	31.7	32.6	33.4	34.1	34.8	35.4	36.0	36.5	37.0	37.5	37.9
3	8.26	10.6	12.2	13.3	14.2	15.0	15.6	16.2	16.7	17.1	17.5	17.9	18.2	18.5	18.8	19.1	19.3	19.5	19.8
4	6.51	8.12	9.17	9.96	10.6	11.1	11.5	11.9	12.3	12.6	12.8	13.1	13.3	13.5	13.7	13.9	14.1	14.2	14.4
5	5.70	6.97	7.80	8.42	8.91	9.32	9.67	9.97	10.24	10.48	10.70	10.89	11.08	11.24	11.40	11.55	11.68	11.81	11.93
6	5.24	6.33	7.03	7.56	7.97	8.32	8.61	8.87	9.10	9.30	9.49	9.65	9.81	9.95	10.08	10.21	10.32	10.43	10.54
7	4.95	5.92	6.54	7.01	7.37	7.68	7.94	8.17	8.37	8.55	8.71	8.86	9.00	9.12	9.24	9.35	9.46	9.55	9.65
8	4.74	5.63	6.20	6.63	6.96	7.24	7.47	7.68	7.87	8.03	8.18	8.31	8.44	8.55	8.66	8.76	8.85	8.94	9.03
9	4.60	5.43	5.96	6.35	6.66	6.91	7.13	7.32	7.49	7.65	7.78	7.91	8.03	8.13	8.23	8.32	8.41	8.49	8.57
10	4.48	5.27	5.77	6.14	6.43	6.67	6.87	7.05	7.21	7.36	7.48	7.60	7.71	7.81	7.91	7.99	8.07	8.15	8.22
11	4.39	5.14	5.62	5.97	6.25	6.48	6.67	6.84	6.99	7.13	7.25	7.36	7.46	7.56	7.65	7.73	7.81	7.88	7.95
12	4.32	5.04	5.50	5.84	6.10	6.32	6.51	6.67	6.81	6.94	7.06	7.17	7.26	7.36	7.44	7.52	7.59	7.66	7.73
13	4.26	4.96	5.40	5.73	5.98	6.19	6.37	6.53	6.67	6.79	6.90	7.01	7.10	7.19	7.27	7.34	7.42	7.48	7.55
14	4.21	4.89	5.32	5.63	5.88	6.08	6.26	6.41	6.54	6.66	6.77	6.87	6.96	7.05	7.12	7.20	7.27	7.33	7.39

15	4.17	4.83	5.25	5.56	5.80	5.99	6.16	6.31	6.44	6.55	6.66	6.76	6.84	6.93	7.00	7.07	7.14	7.20	7.26
16	4.13	4.78	5.19	5.49	5.72	5.92	6.08	6.22	6.35	6.46	6.56	6.66	6.74	6.82	6.90	6.97	7.03	7.09	7.15
17	4.10	4.74	5.14	5.43	5.66	5.85	6.01	6.15	6.27	6.38	6.48	6.57	6.66	6.73	6.80	6.87	6.94	7.00	7.05
18	4.07	4.70	5.09	5.38	5.60	5.79	5.94	6.08	6.20	6.31	6.41	6.50	6.58	6.65	6.72	6.79	6.85	6.91	6.96
19	4.05	4.67	5.05	5.33	5.55	5.73	5.89	6.02	6.14	6.25	6.34	6.43	6.51	6.58	6.65	6.72	6.78	6.84	6.89
20	4.02	4.64	5.02	5.29	5.51	5.69	5.84	5.97	6.09	6.19	6.29	6.37	6.45	6.52	6.59	6.65	6.71	6.76	6.82
24	3.96	4.54	4.91	5.17	5.37	5.54	5.69	5.81	5.92	6.02	6.11	6.19	6.26	6.33	6.39	6.45	6.51	6.56	6.61
30	3.89	4.45	4.80	5.05	5.24	5.40	5.54	5.65	5.76	5.85	5.93	6.01	6.08	6.14	6.20	6.26	6.31	6.36	6.41
40	3.82	4.37	4.70	4.93	5.11	5.27	5.39	5.50	5.60	5.69	5.77	5.84	5.90	5.96	6.02	6.07	6.12	6.17	6.21
60	3.76	4.28	4.60	4.82	4.99	5.13	5.25	5.36	5.45	5.53	5.60	5.67	5.73	5.79	5.84	5.89	5.93	5.98	6.02
120	3.70	4.20	4.50	4.71	4.87	5.01	5.12	5.21	5.30	5.38	5.44	5.51	5.56	5.61	5.66	5.71	5.75	5.79	5.83
∞	3.64	4.12	4.40	4.60	4.76	4.88	4.99	5.08	5.16	5.23	5.29	5.35	5.40	5.45	5.49	5.54	5.57	5.61	5.65

[a] This table is reproduced with the permission of Professor E. S. Pearson from May, J. M., "Extended and corrected tables of the upper percentage points of the studentized range." *Biometrika*, **39**, p. 192, 1952.

Table VIIIA   Critical Values of F at 5% Level

f_2 Denominator	\multicolumn{11}{c}{f_1 Numerator}
	1	2	3	4	5	6	7	8	9	10	12
1	161	200	216	225	230	234	237	239	241	242	244
2	18.5	19.0	19.2	19.2	19.3	19.3	19.4	19.4	19.4	19.4	19.4
3	10.1	9.55	9.28	9.12	9.01	8.94	8.89	8.85	8.81	8.79	8.74
4	7.71	6.94	6.59	6.39	6.26	6.16	6.09	6.04	6.00	5.96	5.91
5	6.61	5.79	5.41	5.19	5.05	4.95	4.88	4.82	4.77	4.74	4.68
6	5.99	5.14	4.76	4.53	4.39	4.28	4.21	4.15	4.10	4.06	4.00
7	5.59	4.74	4.35	4.12	3.97	3.87	3.79	3.73	3.68	3.64	3.57
8	5.32	4.46	4.07	3.84	3.69	3.58	3.50	3.44	3.39	3.35	3.28
9	5.12	4.26	3.86	3.63	3.48	3.37	3.29	3.23	3.18	3.14	3.07
10	4.96	4.10	3.71	3.48	3.33	3.22	3.14	3.07	3.02	2.98	2.91
11	4.84	3.98	3.59	3.36	3.20	3.09	3.01	2.95	2.90	2.85	2.79
12	4.75	3.89	3.49	3.26	3.11	3.00	2.91	2.85	2.80	2.75	2.69
13	4.67	3.81	3.41	3.18	3.03	2.92	2.83	2.77	2.71	2.67	2.60
14	4.60	3.74	3.34	3.11	2.96	2.85	2.76	2.70	2.65	2.60	2.53
15	4.54	3.68	3.29	3.06	2.90	2.79	2.71	2.64	2.59	2.54	2.48

16	4.49	3.63	3.24	3.01	2.85	2.74	2.66	2.59	2.54	2.49	2.42
17	4.45	3.59	3.20	2.96	2.81	2.70	2.61	2.55	2.49	2.45	2.38
18	4.41	3.55	3.16	2.93	2.77	2.66	2.58	2.51	2.46	2.41	2.34
19	4.38	3.52	3.13	2.90	2.74	2.63	2.54	2.48	2.42	2.38	2.31
20	4.35	3.49	3.10	2.87	2.71	2.60	2.51	2.45	2.39	2.35	2.28
21	4.32	3.47	3.07	2.84	2.68	2.57	2.49	2.42	2.37	2.32	2.25
22	4.30	3.44	3.05	2.82	2.66	2.55	2.46	2.40	2.34	2.30	2.23
23	4.28	3.42	3.03	2.80	2.64	2.53	2.44	2.37	2.32	2.27	2.20
24	4.26	3.40	3.01	2.78	2.62	2.51	2.42	2.36	2.30	2.25	2.18
25	4.24	3.39	2.99	2.76	2.60	2.49	2.40	2.34	2.28	2.24	2.16
30	4.17	3.32	2.92	2.69	2.53	2.42	2.33	2.27	2.21	2.16	2.09
40	4.08	3.23	2.84	2.61	2.45	2.34	2.25	2.18	2.12	2.08	2.00
60	4.00	3.15	2.76	2.53	2.37	2.25	2.17	2.10	2.04	1.99	1.92
120	3.92	3.07	2.68	2.45	2.29	2.18	2.09	2.02	1.96	1.91	1.83
∞	3.84	3.00	2.60	2.37	2.21	2.10	2.01	1.94	1.88	1.83	1.75

[a] This table is reproduced with the permission of Professor E. S. Pearson from Merrington, M., and Thompson, C. M., "Tables of percentage points of the inverted beta (F) distribution," *Biometrika*, **33**, p. 73, 1943.

Table VIIIB Critical Values of F at 1% Level[a]

f_2 Denominator	f_1 Numerator										
	1	2	3	4	5	6	7	8	9	10	12
1	4052	5000	5403	5625	5764	5859	5928	5982	6023	6056	6106
2	98.5	99.0	99.2	99.2	99.3	99.3	99.4	99.4	99.4	99.4	99.4
3	34.1	30.8	29.5	28.7	28.2	27.9	27.7	27.5	27.3	27.2	27.1
4	21.2	18.0	16.7	16.0	15.5	15.2	15.0	14.8	14.7	14.5	14.4
5	16.3	13.3	12.1	11.4	11.0	10.7	10.5	10.3	10.2	10.1	9.89
6	13.7	10.9	9.78	9.15	8.75	8.47	8.26	8.10	7.98	7.87	7.72
7	12.2	9.55	8.45	7.85	7.46	7.19	6.99	6.84	6.72	6.62	6.47
8	11.3	8.65	7.59	7.01	6.63	6.37	6.18	6.03	5.91	5.81	5.67
9	10.6	8.02	6.99	6.42	6.06	5.80	5.61	5.47	5.35	5.26	5.11
10	10.0	7.56	6.55	5.99	5.64	5.39	5.20	5.06	4.94	4.85	4.71
11	9.65	7.21	6.22	5.67	5.32	5.07	4.89	4.74	4.63	4.54	4.40
12	9.33	6.93	5.95	5.41	5.06	4.82	4.64	4.50	4.39	4.30	4.16
13	9.07	6.70	5.74	5.21	4.86	4.62	4.44	4.30	4.19	4.10	3.96
14	8.86	6.51	5.56	5.04	4.70	4.46	4.28	4.14	4.03	3.94	3.80
15	8.68	6.36	5.42	4.89	4.56	4.32	4.14	4.00	3.89	3.80	3.67

16	8.53	6.23	5.29	4.77	4.44	4.20	4.03	3.89	3.78	3.69	3.55
17	8.40	6.11	5.19	4.67	4.34	4.10	3.93	3.79	3.68	3.59	3.46
18	8.29	6.01	5.09	4.58	4.25	4.01	3.84	3.71	3.60	3.51	3.37
19	8.19	5.93	5.01	4.50	4.17	3.94	3.77	3.63	3.52	3.43	3.30
20	8.10	5.85	4.94	4.43	4.10	3.87	3.70	3.56	3.46	3.37	3.23
21	8.02	5.78	4.87	4.37	4.04	3.81	3.64	3.51	3.40	3.31	3.17
22	7.95	5.72	4.82	4.31	3.99	3.76	3.59	3.45	3.35	3.26	3.12
23	7.88	5.66	4.76	4.26	3.94	3.71	3.54	3.41	3.30	3.21	3.07
24	7.82	5.61	4.72	4.22	3.90	3.67	3.50	3.36	3.26	3.17	3.03
25	7.77	5.57	4.68	4.18	3.86	3.63	3.46	3.32	3.22	3.13	2.99
30	7.56	5.39	4.51	4.02	3.70	3.47	3.30	3.17	3.07	2.98	2.84
40	7.31	5.18	4.31	3.83	3.51	3.29	3.12	2.99	2.89	2.80	2.66
60	7.08	4.98	4.13	3.65	3.34	3.12	2.95	2.82	2.72	2.65	2.50
120	6.85	4.79	3.95	3.48	3.17	2.96	2.79	2.66	2.56	2.47	2.34
∞	6.63	4.61	3.78	3.32	3.02	2.80	2.64	2.51	2.41	2.32	2.18

[a] This table is reproduced with the permission of Professor E. S. Pearson from Merrington, M., and Thompson, C. M., "Tables of percentage points of the inverted beta (F) distribution." *Biometrika*, **33**, p. 73, 1943.

Table IXA A_2* Factors for Control Charts[a]

k	\(n\) 2	3	4	5	6	7	8	9	10
1			4.30	1.74	1.18	0.954	0.799	0.747	0.612
2		2.66	1.41	1.00	0.788	0.646	0.558	0.487	0.443
3	6.14	1.82	1.15	0.834	0.667	0.563	0.490	0.437	0.392
4	4.57	1.60	1.02	0.760	0.616	0.526	0.468	0.402	0.373
5	3.61	1.48	0.952	0.720	0.582	0.505	0.441	0.392	0.359
6	3.26	1.38	0.915	0.695	0.568	0.484	0.427	0.384	0.351
7	3.04	1.33	0.886	0.677	0.554	0.475	0.421	0.380	0.342
8	2.90	1.29	0.867	0.662	0.545	0.472	0.416	0.375	0.341
9	2.79	1.27	0.853	0.655	0.540	0.466	0.412	0.370	0.336
10	2.69	1.24	0.838	0.647	0.534	0.460	0.407	0.366	0.333
15	2.39	1.18	0.803	0.621	0.520	0.445	0.391	0.350	0.318
20	2.25	1.13	0.786	0.609	0.504	0.430	0.381	0.345	0.316
25	2.14	1.11	0.774	0.602	0.494	0.428	0.379	0.343	0.308
∞	1.88	1.02	0.729	0.577	0.483	0.419	0.373	0.337	0.308

[a] Values given are 3σ limits.

Table IXB D_3^* *Factors for Lower-Range Chart Limits*[a]

k	n								
	2	3	4	5	6	7	8	9	10
1	0.008	0.071	0.143	0.204	0.252	0.289	0.321	0.347	0.369
2	0.008	0.076	0.154	0.219	0.271	0.312	0.344	0.374	0.396
3	0.008	0.077	0.157	0.225	0.278	0.320	0.354	0.383	0.408
4	0.008	0.077	0.159	0.228	0.282	0.324	0.360	0.389	0.414
5	0.008	0.077	0.160	0.230	0.285	0.328	0.364	0.393	0.417
6	0.008	0.078	0.162	0.231	0.286	0.329	0.365	0.395	0.419
7	0.008	0.078	0.163	0.232	0.287	0.331	0.366	0.397	0.422
8	0.008	0.079	0.163	0.233	0.288	0.332	0.368	0.398	0.423
9	0.008	0.079	0.164	0.234	0.289	0.333	0.369	0.399	0.424
10	0.008	0.080	0.164	0.234	0.290	0.334	0.370	0.400	0.425
15	0.008	0.080	0.165	0.236	0.292	0.337	0.374	0.404	0.428
20	0.008	0.080	0.165	0.236	0.293	0.338	0.374	0.405	0.430
25	0.008	0.080	0.166	0.237	0.294	0.339	0.375	0.405	0.431
∞	0.008	0.080	0.166	0.239	0.296	0.341	0.378	0.408	0.434

[a] Values given are 3σ limits.

Table IXC D_4* Factors for Upper-Range Chart Limits[a]

k/n	2	3	4	5	6	7	8	9	10
1		14.1	6.70	5.10	4.16	3.60	3.20	2.99	2.82
2	14.7	5.58	3.87	3.19	2.86	2.58	2.43	2.32	2.24
3	9.35	4.24	3.24	2.76	2.52	2.34	2.21	2.12	2.06
4	7.00	3.76	2.92	2.57	2.37	2.23	2.11	2.04	1.97
5	6.02	3.46	2.79	2.47	2.29	2.16	2.06	1.99	1.92
6	5.39	3.29	2.71	2.40	2.23	2.11	2.02	1.95	1.90
7	5.08	3.20	2.65	2.35	2.19	2.07	1.99	1.93	1.87
8	4.87	3.13	2.60	2.32	2.16	2.05	1.98	1.92	1.86
9	4.64	3.09	2.56	2.30	2.14	2.03	1.96	1.90	1.86
10	4.48	3.05	2.53	2.27	2.12	2.01	1.94	1.89	1.85
15	4.20	2.89	2.44	2.21	2.07	1.97	1.91	1.86	1.82
20	4.02	2.81	2.39	2.18	2.04	1.94	1.89	1.84	1.79
25	3.91	2.76	2.36	2.17	2.04	1.93	1.88	1.83	1.76
∞	3.27	2.58	2.28	2.11	2.00	1.92	1.86	1.82	1.78

[a] Values given are 3σ limits.

Table IXD A_2* Factors for Average Charts

	$\alpha = 0.05$					$\alpha = 0.01$			
	n						*n*		
k	2	3	4	5	k	2	3	4	5
5	1.77	0.821	0.558	0.436	5	2.54	1.17	0.780	0.596
6	1.64	0.793	0.548	0.426	6	2.39	1.12	0.754	0.579
7	1.58	0.775	0.537	0.418	7	2.30	1.08	0.735	0.566
8	1.53	0.764	0.529	0.412	8	2.24	1.06	0.719	0.557
9	1.50	0.759	0.524	0.409	9	2.16	1.04	0.710	0.550
10	1.47	0.754	0.518	0.407	10	2.11	0.992	0.702	0.545
15	1.39	0.722	0.512	0.396	15	1.94	0.977	0.677	0.527
20	1.36	0.707	0.499	0.391	20	1.86	0.952	0.665	0.518
25	1.33	0.702	0.491	0.387	25	1.80	0.937	0.654	0.513
∞	1.23	0.668	0.476	0.377	∞	1.62	0.878	0.626	0.495

Table IXE   D₃* Factors for Lower Control Limits for Range

| | $\alpha = 0.05$ | | | | | $\alpha = 0.01$ | | | |
| | n | | | | | n | | | |
k	2	3	4	5	k	2	3	4	5
5	0.079	0.249	0.358	0.430	5	0.016	0.110	0.203	0.276
6	0.079	0.250	0.360	0.432	6	0.016	0.110	0.205	0.278
7	0.079	0.251	0.362	0.433	7	0.016	0.111	0.206	0.279
8	0.079	0.253	0.363	0.434	8	0.016	0.112	0.206	0.280
9	0.080	0.254	0.364	0.435	9	0.016	0.112	0.208	0.280
10	0.080	0.255	0.365	0.436	10	0.016	0.113	0.208	0.281
15	0.080	0.255	0.367	0.438	15	0.016	0.113	0.209	0.282
20	0.080	0.255	0.368	0.439	20	0.016	0.113	0.210	0.283
25	0.079	0.255	0.368	0.440	25	0.016	0.113	0.210	0.284
∞	0.079	0.255	0.369	0.443	∞	0.016	0.113	0.211	0.286

Table IXF   D₄* Upper Control Limits for Range

	$\alpha = 0.05$					$\alpha = 0.01$			
	n						*n*		
k	2	3	4	5	*k*	2	3	4	5
5	3.14	2.26	1.97	1.81	5	5.01	3.11	2.56	2.27
6	3.01	2.21	1.94	1.78	6	4.61	2.97	2.47	2.22
7	2.91	2.17	1.91	1.76	7	4.37	2.90	2.42	2.18
8	2.86	2.15	1.89	1.75	8	4.22	2.84	2.38	2.16
9	2.83	2.13	1.88	1.74	9	4.13	2.80	2.36	2.14
10	2.81	2.12	1.87	1.73	10	4.03	2.78	2.33	2.12
15	2.69	2.07	1.84	1.71	15	3.77	2.65	2.28	2.07
20	2.63	2.04	1.83	1.70	20	3.61	2.59	2.25	2.05
25	2.62	2.03	1.82	1.69	25	3.48	2.57	2.22	2.03
∞	2.46	1.96	1.76	1.65	∞	3.23	2.43	2.14	1.98

Table X   Critical Values for Discarding Invalid Measurements

n	t_i
3	1.53
4	1.05
5	0.86
6	0.76
7	0.69
8	0.64
9	0.60
10	0.58
11	0.56
12	0.54
13	0.52
14	0.51
15	0.50
20	0.46

The probability is approximately 0.95 that if $t_i = |X - \bar{X}|/R$ is greater than tabulated t_i the value being investigated is invalid.

Table XI   t Distribution[a]

df	α			
	0.1	0.05	0.01	0.001
1	6.314	12.71	63.66	636.6
2	2.920	4.303	9.925	31.60
3	2.353	3.182	5.841	12.94
4	2.132	2.776	4.604	8.610
5	2.015	2.571	4.032	6.859
6	1.943	2.447	3.707	5.959
7	1.895	2.365	3.499	5.405
8	1.860	2.306	3.355	5.041
9	1.833	2.262	3.250	4.781
10	1.812	2.228	3.169	4.587
11	1.796	2.201	3.106	4.437
12	1.782	2.179	3.055	4.318
13	1.771	2.160	3.012	4.221
14	1.761	2.145	2.977	4.140
15	1.753	2.131	2.947	4.073

Table XI—continued.

df	α			
	0.1	0.05	0.01	0.001
16	1.746	2.120	2.921	4.015
17	1.740	2.110	2.898	3.965
18	1.734	2.101	2.878	3.922
19	1.729	2.093	2.861	3.883
20	1.725	2.086	2.845	3.850
21	1.721	2.080	2.831	3.819
22	1.717	2.074	2.819	3.792
23	1.714	2.069	2.807	3.767
24	1.711	2.064	2.797	3.745
25	1.708	2.060	2.787	3.725
26	1.706	2.056	2.779	3.707
27	1.703	2.052	2.771	3.690
28	1.701	2.048	2.763	3.674
29	1.699	2.045	2.756	3.659
30	1.697	2.042	2.750	3.646
∞	1.645	1.960	2.576	3.291
(one-tail)				
α	0.05	0.025	0.005	0.0005

[a] Values of t that will be exceeded with a probability α for df degrees of freedom (two-tail).

Index

A

Accuracy, 6–7, 149–150
Analysis of variance, 27–39
  basic assumptions of, 31–34
  models of, 42
  by range, 71–93
Analytical errors, 2, 13, 160
Attribute sampling, 136–137
Average, 2, 4, 13
  control chart for, 98, 101–108
Average deviation, 7

B

Bias, 2, 109, 110, 111, 160
Block design classification
  one-way, 32–36, 72–74
  two-way, 36–39, 75–77

C

Colorimetric analysis, 120–126
  precision of, 152
Compliance with specifications, 156–157
Components of variance, 42–44, 86–88, 91–92, 139–141
  sample size and, 139–141
Confidence limits, 14–19, 23

Constant error, 109
Control chart, 95–108
  analytical lab, 157
  averages, 98, 101–108
  examples, 101–108
  nomenclature of, 95–96
  ranges, 98
  runs and trends in, 100
  theory of, 96–97
Control limits, 97
  calculation of, 99
  significance of, 100
Correlated variables, 109–130

D

Degrees of freedom, 9, 35, 38, 57
Determinate errors, 2

E

Equivalent degrees of freedom, 71
Error
  analytical, 2, 13, 160
  constant, 2, 110
  determinate, 2
  experimental, 2, 13, 32, 33, 99, 145
  indeterminate, 2
  random, 14, 162
Expected mean square, 42–44

191

Expected mean square components, 44–46
Experimental design, 27, 29, *see also* names of specific designs
 criteria, 28–29
Experimental error, 2, 28, 32, 33, 99, 145

F

F Ratio, 31, 35, 43
 table, 178–181
Factorial experiment, 49–54, 84–89
 nested, 55–59, 89–92
Fraction defective, 134, 135

H

Hypothesis, 27
Hypothesis, null, 30

I

Indeterminate errors, 2
Interaction, 28, 39, 77
Intercept, 109, 113
 short-cut methods, 116
Interlaboratory studies, 158–163
 analysis, 158
 design, 158
 Youden's method, 159
Invalid measurements, 22–23, 146–147
 table, 188

L

L Test, 67, 68, 120, 121, 123, 150
 table, 167
Lack of control, 106–108
Latin square, 47–49, 80–84
Least squares, 110
Linear regression, 109–110

M

M Test, 67, 68, 149
 table, 168–170
MIL–STD–105, 136–137
MIL–STD–414, 138
Models of analysis of variance, 42

N

Nonlinear regression, 127–130
Normal distribution, 3–4, 97
 reduced sampling based on, 153–157
 sampling plans based on, 136–137, 141–144
Null hypothesis, 30
Numbers, 1

O

Operating characteristic curve, **134**, 138
Optical rotation (precision of), 151–**152**
Outliers, 22–23, 146, 147
 table, 188

P

Precision, 6–7, 113–114, 147–148, 151, 152
Probability, 3

Q

q Test, 71, 73, 77, 88
 table, 174–177

R

Random sample, 133
Range, 10–11

Range—*cont.*
  analysis of variance by, 71–93
  confidence limits calculated from, 18, 19
    table, 165
  control chart for, 98
  control limits calculated from, 99–100
    tables, 182–187
  factors for calculations, 23–24
  relationship to Standard Deviation, 11
  studentized, 71
    table, 174–177
Reduced sample size, 153
Regression, 109–130
  laboratory use of, 110
  linear, 109
  nonlinear, 127–130
  short-cut methods, 116–126
Replication, 13–14, 15
Routine analysis, 145–163
Risks, 15, 133, 135, 137
Runs and trends, 100

S

Sample and population, 131–133
Sample size, 135
Sampling, 131–145
  attribute, 136–137
  risks, 133, 135, 137
  theory of, 133–134
  variables, 137–138, 141

Sampling plans
  AOQL, 136
  AQL, 136
  LTPD, 136
  MIL–STD–105D, 136
  MIL–STD–414, 138
Slope, 109, 111, 117
  short-cut methods, 116
Specifications, 156–157
Standard deviation
  calculated from range, 11, 23, 71
  of population, 4
  of sample, 5, 8–9, 11, 17, 20, 113–114
Statistical proof, 30
Student distribution, 5
Student's *t*, 5
Studentized range, 71

T

t Test, 61–67, 117, 149, 150
  average versus standard, 62
  paired comparison, 65
  substitute, 67–70, 121, 149, 150
    table, 189
  unpaired comparison, 63
Tests of significance, 5, 30, *see also* names of specific tests
Tolerance limits, 21–22, 24

V

Variance, 8–10, *see also* Analysis of variance
  components of, 42–44, 86–88, 91–92, 139–141